SpringerBriefs in Statistics

For further volumes:
http://www.springer.com/series/8921

Thomas W. MacFarland

Introduction to Data Analysis and Graphical Presentation in Biostatistics with R

Statistics in the Large

Thomas W. MacFarland
Office for Institutional
  Effectiveness
Nova Southeastern University
Fort Lauderdale, FL, USA

ISSN 2191-544X   ISSN 2191-5458 (electronic)
ISBN 978-3-319-02531-5   ISBN 978-3-319-02532-2 (eBook)
DOI 10.1007/978-3-319-02532-2
Springer Cham Heidelberg New York Dordrecht London

Library of Congress Control Number: 2013953880

© The Author(s) 2014
This work is subject to copyright. All rights are reserved by the Publisher, whether the whole or part of the material is concerned, specifically the rights of translation, reprinting, reuse of illustrations, recitation, broadcasting, reproduction on microfilms or in any other physical way, and transmission or information storage and retrieval, electronic adaptation, computer software, or by similar or dissimilar methodology now known or hereafter developed. Exempted from this legal reservation are brief excerpts in connection with reviews or scholarly analysis or material supplied specifically for the purpose of being entered and executed on a computer system, for exclusive use by the purchaser of the work. Duplication of this publication or parts thereof is permitted only under the provisions of the Copyright Law of the Publisher's location, in its current version, and permission for use must always be obtained from Springer. Permissions for use may be obtained through RightsLink at the Copyright Clearance Center. Violations are liable to prosecution under the respective Copyright Law.
The use of general descriptive names, registered names, trademarks, service marks, etc. in this publication does not imply, even in the absence of a specific statement, that such names are exempt from the relevant protective laws and regulations and therefore free for general use.
While the advice and information in this book are believed to be true and accurate at the date of publication, neither the authors nor the editors nor the publisher can accept any legal responsibility for any errors or omissions that may be made. The publisher makes no warranty, express or implied, with respect to the material contained herein.

Printed on acid-free paper

Springer is part of Springer Science+Business Media (www.springer.com)

Contents

1	**Introduction: Biostatistics and R**		1
	1.1 Purpose of This Text		1
	1.2 Development of Biostatistics		2
	1.3 Development of R		3
	1.4 How R is Used in This Text		4
2	**Data Exploration, Descriptive Statistics, and Measures of Central Tendency**		5
	2.1 Background on This Lesson		5
		2.1.1 Description of the Data	5
		2.1.2 Null Hypothesis (Ho)	7
	2.2 Data Import of a .csv Spreadsheet-Type Data File into R		7
	2.3 Organize the Data and Display the Code Book		9
	2.4 Conduct a Visual Data Check		9
	2.5 Descriptive Analysis of the Data		10
	2.6 Summary		13
	2.7 Addendum: Specialized External Packages and Functions		13
	2.8 Prepare to Exit, Save, and Later Retrieve This R Session		15
3	**Student's t-Test for Independent Samples**		17
	3.1 Background on This Lesson		17
		3.1.1 Description of the Data	17
		3.1.2 Null Hypothesis (Ho)	18
	3.2 Data Import of a .csv Spreadsheet-Type Data File into R		19
	3.3 Organize the Data and Display the Code Book		20
	3.4 Conduct a Visual Data Check		23
	3.5 Descriptive Analysis of the Data		34
	3.6 Conduct the Statistical Analysis		40
	3.7 Summary		42
	3.8 Addendum: t-Statistic v z-Statistic		43
		3.8.1 Create the Enumerated Dataset	44

		3.8.2 Calculate the t-Statistic	44
		3.8.3 Calculate the z-Statistic	45
	3.9	Prepare to Exit, Save, and Later Retrieve This R Session	45
4	**Student's t-Test for Matched Pairs**		**47**
	4.1	Background on This Lesson	47
		4.1.1 Description of the Data	47
		4.1.2 Null Hypothesis (Ho)	49
		4.1.3 Unstacked Data and Stacked Data	49
	4.2	Data Import of a .csv Spreadsheet-Type Data File into R	51
	4.3	Organize the Data and Display the Code Book	52
	4.4	Conduct a Visual Data Check	54
	4.5	Descriptive Analysis of the Data	60
	4.6	Conduct the Statistical Analysis	63
	4.7	Summary	65
	4.8	Addendum 1: Stacked Data and Student's t-Test for Matched Pairs	66
	4.9	Addendum 2: The Impact of N on Student's t-Test	70
	4.10	Prepare to Exit, Save, and Later Retrieve This R Session	72
5	**Oneway Analysis of Variance (ANOVA)**		**73**
	5.1	Background on This Lesson	73
		5.1.1 Description of the Data	73
		5.1.2 Null Hypothesis (Ho)	75
	5.2	Data Import of a .csv Spreadsheet-Type Data File into R	75
	5.3	Organize the Data and Display the Code Book	77
	5.4	Conduct a Visual Data Check	82
	5.5	Descriptive Analysis of the Data	87
	5.6	Conduct the Statistical Analysis	89
		5.6.1 Exploratory Oneway ANOVA	90
		5.6.2 Oneway ANOVA Method 1: lm() and anova() Functions	91
		5.6.3 Oneway ANOVA Method 2: aov() and TukeyHSD() Functions	92
	5.7	Summary	93
	5.8	Addendum: Other Packages for Display of Oneway ANOVA	96
	5.9	Prepare to Exit, Save, and Later Retrieve This R Session	97
6	**Twoway Analysis of Variance (ANOVA)**		**99**
	6.1	Background on This Lesson	99
		6.1.1 Description of the Data	99
		6.1.2 Null Hypothesis (Ho)	100
	6.2	Data Import of a .csv Spreadsheet-Type Data File into R	100
	6.3	Organize the Data and Display the Code Book	101
	6.4	Conduct a Visual Data Check	104
	6.5	Descriptive Analysis of the Data	111
	6.6	Conduct the Statistical Analysis	117

	6.7	Summary	122
	6.8	Addendum: Other Packages for Display of Twoway ANOVA	124
	6.9	Prepare to Exit, Save, and Later Retrieve This R Session	126
7	**Correlation and Linear Regression**		**129**
	7.1	Background on This Lesson	129
		7.1.1 Description of the Data	129
		7.1.2 Null Hypothesis (Ho)	130
	7.2	Data Import of a .csv Spreadsheet-Type Data File into R	131
	7.3	Organize the Data and Display the Code Book	132
	7.4	Conduct a Visual Data Check	135
	7.5	Descriptive Analysis of the Data	140
	7.6	Conduct the Statistical Analysis	142
		7.6.1 Correlation Using Pearson's r	142
		7.6.2 Linear Regression	150
	7.7	Summary	154
	7.8	Addendum: Multiple Regression	155
		7.8.1 Hand-Calculate Multiple Regression	156
		7.8.2 Minimal Adequate Model (MAM) for Regression	158
		7.8.3 Stepwise Regression	160
	7.9	Prepare to Exit, Save, and Later Retrieve This R Session	163
8	**Future Actions and Next Steps**		**165**
	8.1	Use of This Text	165
	8.2	Future Use of R for Biostatistics	166
	8.3	External Resources	167
	8.4	Contact the Author	167

Chapter 1
Introduction: Biostatistics and R

Abstract The purpose of this lesson is to provide context for the science of biostatistics and to highlight a few of the major contributors. Emphasis is given to the role of data analysis for the various disciplines in the biological sciences (e.g., agriculture, biology, clinical trials, ecology, environmental health, epidemiology, genetics, health sciences, nutrition, public health, etc.). The practice of biostatistics is then linked to the use of R, a free and open source software environment. As explained, each problem in this text is associated with a .csv (comma-separated values) ASCII file, a Code Book detailing data organization, quality assurance through graphical presentations and descriptive statistics, selected statistical analyses, summary of outcomes, and an addendum offering ideas on how R can be used for additional insight into biostatistics.

Keywords Agriculture • Biology • Biostatistics • Census • Clinical trials • Code Book • Comma-separated values ASCII file • Command Line Interface (CLI) • Comprehensive R Archive Network (CRAN) • CRAN Contributed Packages • Data analysis • Descriptive statistics • Ecology • Environmental health • Epidemiology • Genetics • Graphical User Interface (GUI) • Health sciences • Nutrition • Open source software • Public health • R • S • Scheme

1.1 Purpose of This Text

Scientists use empiricism to guide and validate decisions. Precision, orderliness, analysis, and a sound background in statistics are directly associated with informed judgment, decision-making, and the subsequent allocation of human, physical, and fiscal resources – all to improve the human condition. The purpose of this text is to provide an introduction to the use of R software as a platform for problems related to biostatistics. Data identification, data organization, graphical and descriptive portrayal of phenomena, and statistical tests through the use of R are all inherent to this text.

R supports a Graphical User Interface (GUI), the R Commander. This resource is available as an external R package, Rcmdr. Rcmdr is fairly easy to use but eventually there are limits on the use of R Commander.

R also supports a far more robust and useful syntax-based Command Line Interface (CLI) approach to statistics. This text is focused on the use of R-based syntax, working at the command line, to address data organization, statistical analyses, and graphical presentations as they relate to biostatistics. A series of small confidence-building activities are presented at the beginning of this text, with more detail gradually introduced as the text is followed from beginning to end. All examples are for biostatistics. The many examples presented in this text can be easily applied to all areas of biostatistics, regardless of major area of study.

1.2 Development of Biostatistics

The term statistics is derived from *status*, the Latin term for state. Thus, the science and practice of statistics, as we think of it today, was first associated with data relating to the state, such as census counts and health records. Given the importance of statistics as a part of state governance, there are more than a few accounts of census-taking and health records from the earliest days of recorded history.

Going beyond mere record-keeping, an interest in the mathematics of chance (e.g., probability) began to develop in the 1500s and 1600s, especially among those who engaged in European court life. The early interest in probability may not have been altruistic but was instead focused on gaining advantage in card games and other forms of gambling. The use of probability to solve problems for societal gain may not have been the first interest but instead attention was focused on the question, *Given that there a X cards in the deck, if I discard the Y card from my hand, what is the chance that I will draw the Z card from the deck and improve my chance of winning this game of cards?*

This early interest in probability and eventually the evolving science of statistics as a vehicle for social improvement eventually grew into what we think of as biostatistics. It is far beyond the purpose of this introductory text on the use of R in biostatistics to go into too much detail, but at a minimum it would be helpful to look into the biography and contributions of the following founders of what we now consider biostatistics:

- **Blaise Pascal** (1623–1662), prepared early writings on probability and developed the Pascaline (e.g., mechanical calculator).
- **John Graunt** (1620–1674), published *Natural and Political Observations Made Upon the Bills of Mortality*, perhaps the first widely-read text on demographics, public health, and epidemiology.
- **John Snow** (1813–1858), advocated for epidemiology and the 1854 Broad Street (London) Cholera Outbreak.

- **Florence Nightingale** (1820–1910), although perhaps best known as an advocate for our modern view of nursing, *Diagram of the Causes of Mortality in the Army in the East* was a breakthrough publication that had strong implications for how biostatistics could be used to improve public health.
- **Ronald Fisher** (1890–1962), the publication *Statistical Methods for Research Workers* and other works are still central to how data are used in biostatistics.

Although Fisher may be the immediate answer if anyone were asked to identify a famous biostatistician, Snow should also be singled out. To put the many individuals who contributed to our current view of biostatistics into context, consider Snow's work during the mid-1850s London cholera (e.g., *Vibrio cholerae*) outbreak and his then innovative use of mapping techniques based on data gained through exhaustive empirical methods. Far from being an academic who dealt only in theory, Snow put his own life at risk to obtain the data needed to validate that cholera was a waterborne pathogen. Then, he used persuasive argumentation with public officials, based on scientific outcomes, to confront the problem and take appropriate actions.

1.3 Development of R

R was first developed in the early-to-mid 1990s, drawing from programming features previously used with S and Scheme. R provides an excellent environment for the organization, statistical analysis, and graphical presentation of data. As opposed to the well-known proprietary statistical analysis software programs, R is both open source and free to download.

R is available through the Comprehensive R Archive Network (CRAN, http://cran.us.r-project.org/). R supports all major operating systems: Linux, Mac, UNIX, and Windows. Again, R is open source software and there is no direct cost for this freely-available software.

The R environment is based on a set number of functions available in the package initially downloaded. The download takes about 10–15 min, depending on speed of Internet connectivity. Then, additional functions are available in external packages. There are currently more than 3,000 external packages hosted through CRAN.

In the nearly 20 years since R was first developed the R community has grown substantially. R has active Internet discussion groups and the R community also supports an annual international conference, typically rotating between Europe and North America.

It cannot be overstated that R is gaining international recognition as a preferred medium for data organization, statistical analysis, and graphical presentation. Quite simply, the free nature of open source software is appealing and the far-reaching use of R is displayed in the many CRAN mirror sites that host R, currently ranging in alphabetical order from Argentina to Vietnam.

1.4 How R is Used in This Text

Each biologically-oriented problem addressed in this text is approached in the same manner, to promote consistency, modularity, and ease of reuse:

- All data are prepared in a .csv (comma-separated values) spreadsheet-type ASCII file format and the data are then imported into R. The various .csv datasets accompany the Web-based resource associated with this text.
- A Code Book is used to communicate data organization and when needed, data are organized into needed format.
- Graphics are used to present a visual data check.
- Descriptive statistics are further used to obtain a better understanding of the data.
- The needed statistical analyses are conducted.
- A summary of outcomes is presented.
- An addendum is used to provide additional insight into the selected test and options on how to enhance the use of R for each statistical test.

Again, small and easy-to-follow confidence-building examples are used at the beginning of this text. Greater complexity is gradually introduced until the final chapters in this text present the use of R in a fairly robust manner.

Chapter 2
Data Exploration, Descriptive Statistics, and Measures of Central Tendency

Abstract The purpose of this lesson is to give attention to descriptive analysis, measures of central tendency, and graphical presentation of data, which are essential before any statistical analyses are conducted. Initial efforts should be placed on data exploration and specifically the use of descriptive statistics and measures of central tendency (e.g., mode, median, mean, standard deviation, etc.). A complete summary of descriptive statistics is presented in this lesson, both for factor-type object variables as well as numeric object variables of an interval or continuous nature. An initial summary of graphical presentations available through R is provided, with emphasis on publishable quality graphics deferred until later lessons.

Keywords Barplot • Boxplot (box-and-whiskers plot) • Boxplot statistics • Data exploration • Density plot • Descriptive statistics • Dotchart • Histogram • Interquartile range (IQR) • Length • Maximum • Maximum location • Mean • Measures of central tendency • Median • Minimum • Minimum location • Mode • Quantile-quantile plot • Quartiles • Range • Scatter plot • Sort • Standard deviation • Stem-and-leaf plot • Stripchart • Sum • Summary • Tukey's five number summary • Variance

2.1 Background on This Lesson

2.1.1 Description of the Data

This lesson on descriptive statistics and measures of central tendency is taken from a study that was conducted at a large high school in Florida, as part of a general investigation of wellness and student health. The dataset for this lesson is fairly small (N = 30 subjects) and represents only a small part of a much larger dataset, larger in terms of more subjects and larger in terms of more variables. This lesson describes the use of R for descriptive statistics and measures of central tendency, with outcomes presented as numerical statistics and simple graphical presentations.

For this lesson, consider the data gained by a school nurse who weighed all 30 students in Computer Programming III (Course Number 0201320), a Computer Science Education course offered to Grade 12 (e.g., High School Seniors, usually 17–18 years old) students. Weight was measured in pounds, with accuracy at the tenth of a pound. As the principal investigator, the school nurse is naturally concerned with overall trends as well as individual measures. What was the average weight? What was the lowest weight and what was the highest weight? What was the variance in weight? Were there any trends that need attention, either for immediate purposes or in the future? With proper analysis, this information could be used, in part, as the basis for informed decision-making on wellness, food selections in the cafeteria, policy and procedures for snack vending machines, etc.

This lesson provides an introduction, using a small sample of only 30 subjects, of how descriptive statistics and measures of central tendency have value on their own and also as indicators for the use of other statistical tests. Quite often when examining data and relationships between and among data, it is useful to offer a general view of the data. Saying this, consider the data conceivably associated with this lesson. It would be more than somewhat useful to know:

- How many students were enrolled in the class and are eligible to have their weights measured?
- How many students had their weights measured?
- What is the average weight and are there multiple definitions of the term average? If there are multiple definitions for the term average, when is it appropriate to use one view of the term average but not the other(s)?
- Did most weights cluster around the average weight, or was there a wide degree of variance in weights?
- Were there any weights that seem to be exceptionally out-of-range (e.g., outliers), demanding specific attention for these observed weights?
- Were there any weights that seem to be illogical, perhaps by accidental data entry of alphabetical characters or similar errors in an object that has otherwise been declared as a vector of numeric values?
- What was the range of weights, from the lowest (e.g., minimum) weight to the highest (e.g., maximum) weight?
- Do the weights display normal distribution, approximating a bell-shaped curve, or is the distribution skewed and if so, how? Are weights skewed to the left or are weights skewed to the right?

Descriptive statistics and measures of central tendency, or representation of the average:

- **Mode**: most frequent measure (An oddity of R is that the mode() function has nothing to do with measures of central tendency, but there are convenient work-arounds that provide mode as an average.)

- **Median**: mid-point of an array of measures
- **Mean**: arithmetic average (Sum/N)

In the perfect bell-shaped curve, all three measures for average (e.g., mode, median, and mean) would be equivalent, but of course this level of perfection is rarely achieved.

Measures of dispersion, spread, or variance in range away from the average:

- **Variance**: the sum of squared deviations from the mean
- **SD**: the standard deviation, or the square root of the variance
- **Range**: the spread from the lowest measure to the highest measure

It is common to present in summary statistics a listing of these descriptive statistics, to give the reader a general view of the data. It is also highly desirable to provide graphical figures, visually representing trends.

This lesson has been designed as a demonstration of how R can be used to provide descriptive statistics and measures of central tendency. The emphasis will be on the use of functions found in the basic R package as well as a brief introduction to the use of functions gained from external R packages. Complementary graphical representations are also provided.

This lesson should provide a fairly detailed introduction to descriptive statistics and measures of central tendency and how they are calculated and presented using R. This topic is of special importance since nearly each statistical analysis associated with parametric data (e.g., the use of interval or ratio data for Student's t-Test, Analysis of Variance, etc.) begins with descriptive statistics and measures of central tendency.

2.1.2 Null Hypothesis (Ho)

Because this lesson is specific only to descriptive statistics, there is no associated Null Hypothesis. The Null Hypothesis will be identified, however, in future lessons.

2.2 Data Import of a .csv Spreadsheet-Type Data File into R

The data for this lesson are from a much larger dataset. The complete dataset was originally prepared in Gnumeric, an open source spreadsheet. After a set of manipulations (largely **Copy and Paste** and later **File and Save as**) the dataset for this lesson was put into .csv (e.g., comma-separated values) file format. The data are in ASCII format and they are separated by commas. The data are not separated by tabs and the data are not separated by spaces.

Eventually, the data were placed on an external harddrive (the F drive) in a directory marked as `R_Biostatistics`. All analyses and presentations start here.

2 Data Exploration, Descriptive Statistics, and Measures of Central Tendency

From this starting point, note below how R is set to work in the appropriate directory and then how the read.table() function is used to read in the comma-separated values .csv format ASCII file that contains the data.

```
################################################################
# Housekeeping                           Use for All Analyses
################################################################
rm(list = ls())      # CAUTION: Remove all files in the working
                     # directory. If this action is not desired,
                     # use the rm() function one-by-one to remove
                     # the objects that are not needed.
setwd("F:/R_Biostatistics")
                     # Set to a new working directory.
                     # Note the single forward slash and double
                     # quotes.
                     # This new directory should be the directory
                     # where the data file is located, otherwise
                     # the data file will not be found.
getwd()              # Confirm the working directory.
search()             # Attached packages and objects.
################################################################
```

Create an object called WeightG12Stu.df. The object WeightG12Stu.df will be a dataframe, as indicated by the enumerated .df extension to the object name. This object will represent the output of applying the read.table() function against the comma-separated values file called WeightGrade12Students.csv. Note the arguments used with the read.table() function, showing that there is a header with descriptive variable names (header = TRUE) and that the separator between fields is a comma (sep = ",").

```
WeightG12Stu.df <- read.table (file =
  "WeightGrade12Students.csv",
  header = TRUE,
  sep = ",")                   # Import the .csv file

getwd()                        # Identify the working directory
ls()                           # List objects
attach(WeightG12Stu.df)        # Attach the data, for later use
str(WeightG12Stu.df)           # Identify structure
nrow(WeightG12Stu.df)          # List the number of rows
ncol(WeightG12Stu.df)          # List the number of columns
dim(WeightG12Stu.df)           # Dimensions of the data frame
names(WeightG12Stu.df)         # Identify names
colnames(WeightG12Stu.df)      # Show column names
rownames(WeightG12Stu.df)      # Show row names
head(WeightG12Stu.df)          # Show the head
tail(WeightG12Stu.df)          # Show the tail
WeightG12Stu.df                # Show the entire dataframe
summary(WeightG12Stu.df)       # Summary statistics
```

2.3 Organize the Data and Display the Code Book

The dataframe WeightG12Stu.df is fairly simple and very little, if anything, needs to be done to organize the data. That will not be the case in later lessons, but this lesson was designed to serve as an easy-to-follow confidence-building introduction to R so in turn a simple dataset was selected for this lesson.

For this simple lesson, the class() function, str() function, and duplicated() function will be sufficient first steps to be sure that data are organized as desired.

```
class(WeightG12Stu.df)
class(WeightG12Stu.df$Subject)  # DataFrame$ObjectName notation
class(WeightG12Stu.df$Weight)   # DataFrame$ObjectName notation

str(WeightG12Stu.df)                    # Structure

duplicated(WeightG12Stu.df$Subject)     # Duplicates
```

The class for each object seems to be correct and there are no duplicate subjects in the sample. A Code Book will help with future understanding of this dataset, even if the data currently seem simple and obvious.

```
############################################################
# Code Book                                                #
############################################################
#                                                          #
# Subject ..................... Factor (e.g. nominal)      #
#              A unique ID ranging from N0000 to N9999     #
#                                                          #
# Weight ................... Numeric (e.g., interval)      #
#                 Weight (tenth of a pound) of Grade 12    #
#                (approximately 17-18 years) high school   #
#                                                 students #
############################################################
```

Labels and recoding of individual object variables are not needed for this simple dataset. However, these actions will be seen in future lessons. Again, small confidence-building activities with easy-to-follow examples are used at the beginning of this set of lessons, with more complexity introduced gradually.

2.4 Conduct a Visual Data Check

As desirable as numeric descriptive statistics and measures of central tendency may be and are therefore often our first thought, to have a full understanding of the data it is necessary to generate graphics, to actually see how data are organized. Graphics provide an essential complement to our understanding of the data. In later lessons other graphics will be demonstrated, but for initial purposes the graphical functions of primary interest are hist(), plot() and plot(density()), boxplot(), stem(), stripchart(), dotchart(), and qqnorm(). Many arguments are available, to embellish these graphical figures, but for now the figures will be prepared in simple format.

The par(ask=TRUE) function and argument are used to freeze the presentation on the screen, one figure at a time. Note how the top line of the figure, under **File - Save as**, provides a variety of graphical formats to save each figure: Metafile, Postscript, PDF, PNG, BMP, TIFF, and JPEG. It is also possible to perform a simple copy and paste against each graphical image. It is also possible to save a graphical image by using R syntax.

```
par(ask=TRUE)
hist(WeightG12Stu.df$Weight)            # Histogram

par(ask=TRUE)
plot(WeightG12Stu.df$Weight)            # Plot

par(ask=TRUE)
plot(density(WeightG12Stu.df$Weight))   # Density plot

par(ask=TRUE)
boxplot(WeightG12Stu.df$Weight)         # Boxplot

stem(WeightG12Stu.df$Weight)            # Stem-and-leaf plot

par(ask=TRUE)
stripchart(WeightG12Stu.df$Weight)      # Stripchat

par(ask=TRUE)
dotchart(WeightG12Stu.df$Weight)        # Dotchart

par(ask=TRUE)
qqnorm(WeightG12Stu.df$Weight)          # Quantile-Quantile plot
```

Again, these initial graphics are simple and currently have no meaningful embellishments. They only serve as a first guide to general trends in data organization. Embellishments to the graphics will be introduced in later lessons, by demonstrating the many arguments used to present titles, prepare text and lines in bold and color, etc.

2.5 Descriptive Analysis of the Data

A series of functions that come with the base R software at initial download can be used to calculate a wide variety of descriptive statistics and measures of central tendency, such as length(), is.na(), complete.cases(), summary(), mean(), sd(), var(), median(), etc. A glaring omission is that the mode() function does not determine the most frequently occurring value but instead provides information on the storage

2.5 Descriptive Analysis of the Data

mode for a R-based object. A specialized function, found in an external R-based package will be used to calculate mode, when mode is viewed as one of three representations of average: mode, median, and mean.

Be sure to notice the DataFrame$ObjectName notation, or WeightG12Stu.df$Weight in this case. This type of specificity calls for a degree of strong typing, but it is a desirable practice and provides protection against unintended naming outcomes.

To learn more about the nature of each R function, use the built-in help features found in R. At the R prompt, key help(function.name) (e.g., help(length), help(summary), help(mean), etc.) to learn the exact nature of each R function.

```
length(WeightG12Stu.df$Weight)
  # Length or N of a vector

is.na(WeightG12Stu.df$Weight)
  # Returns TRUE if indexed value is missing (e.g., NA) and
  # FALSE if indexed value is not missing

complete.cases(WeightG12Stu.df$Weight)
  # Returns TRUE if indexed value is not missing (e.g., NA)
  # and FALSE if indexed value is missing

summary(WeightG12Stu.df$Weight)
  # Descriptive statistics, including NAs if any
```

Output from this simple application of the summary() function follows. The output is basic and in many cases this information is more than sufficient to make judgment on data organization and quality assurance issues.

```
> summary(WeightG12Stu.df$Weight)
   Min. 1st Qu.  Median    Mean 3rd Qu.    Max.
   94.4   117.5   124.6   123.4   130.2   151.5
>    # Descriptive statistics, including NAs if any
```

Other functions for descriptive statistics have value, however, and a few of the many functions associated with descriptive statistics and measures of central tendency are demonstrated below.

```
mean(WeightG12Stu.df$Weight)
  # Mean or arithmetic average

sd(WeightG12Stu.df$Weight)
  # Standard Deviation

var(WeightG12Stu.df$Weight)
  # Variance

median(WeightG12Stu.df$Weight)
  # Median or midpoint
range(WeightG12Stu.df$Weight)
  # Range, minimum and maximum
```

```
min(WeightG12Stu.df$Weight)
  # Minimum

which.min(WeightG12Stu.df$Weight)
  # Location (e.g., index) of the first occurrence of the
  # minimum value

max(WeightG12Stu.df$Weight)
  # Maximum

which.max(WeightG12Stu.df$Weight)
  # Location (e.g., index) of the first occurrence of the
  # maximum value

quantile(WeightG12Stu.df$Weight)
  # Quantiles, or values at:   0%, 25%, 50% 75%, and 100%

sort(WeightG12Stu.df$Weight)
  # Sort or order values in a vector

sum(WeightG12Stu.df$Weight)
  # Arithematic sum of all values in a vector

boxplot.stats(WeightG12Stu.df$Weight)
  # Produce values for a vector related to a boxplot:
  # lower whisker, lower hinge, median, upper hinge, upper
  # whisker, N, and outliers

fivenum(WeightG12Stu.df$Weight)
  # Tukey's five number summary for a vector:  minimum,
  # lower-hinge, median, upper-hinge, and maximum

IQR(WeightG12Stu.df$Weight)
  # Interquartile range of a vector (e.g., a measure of
  # dispersion that is equal to the difference between the
  # upper quartile and the lower quartile

table(WeightG12Stu.df$Weight)
  # Contingency table (e.g., crosstab) of counts for each
  # combination of vector values v factor levels
```

2.6 Summary

Based on the descriptive statistics associated with this lesson, it is evident that the typical student in Computer Programming III weighs approximately 124 pounds, but of course there is variance in weight:

```
N  . . . . . . . . . . . . .    30
Missing . . . . . . . .          0
Median  . . . . . . . . .      124.6
Mean  . . . . . . . . . .      123.3533
SD  . . . . . . . . . . . .     12.90337
Minimum  . . . . . . . .        94.4
Maximum  . . . . . . . .       151.5
```

A review of the histogram and density plot provides assurance that there is fairly normal distribution of weights, following a broad approximation of the bell-shaped curve. Further, the boxplot.stats() function indicated the presence of outliers, both for the minimum weight and the maximum weight. A diligent researcher would look more closely at the outliers, to be sure that the outlier-specific data are correct and that these data do not represent an error in either measurement or data entry.

Quite simply, the descriptive statistics and measures of central tendency for this sample of 30 subjects follows along with useful outcomes and given the approximation of normal distribution of weights, there should be a fair degree of confidence that the students in this sample could be used for other analyses from the larger dataset for any statistical tests that demand normal distribution.

2.7 Addendum: Specialized External Packages and Functions

To use the somewhat humorous expression from a set of well-known American television commercials, *But wait! There's more!* To be specific, there are possibly more than 3,000 external R-based packages available. From these packages there are thousands of specialized functions to supplement the set of functions available when the base R software is initially downloaded. A few specialized functions specific to descriptive statistics and measures of central tendency are demonstrated below.

Be sure to notice how some specialized functions provide not only numerical statistics of immediate use but they also provide a graphical image, to further reinforce the organization of data in question. Function arguments are typically used to embellish graphical output, but in this lesson, function arguments have only been used to any meaningful degree to embellish output from the epicalc::summ() function, to provide a glimpse of potentials that will be enhanced in future lessons.

```
install.packages("asbio")
library(asbio)              # Load the asbio package.
help(package=asbio)         # Show the information page.
sessionInfo()               # Confirm all attached packages.

asbio::Mode(WeightG12Stu.df$Weight)
# Mode, as average (e.g., mode, median, and mean) and not as
# storage mode

> asbio::Mode(WeightG12Stu.df$Weight)
[1] 120.9
> # Mode, as average (e.g., mode, median, and mean) and not as
> # storage mode

install.packages("lessR")
library(lessR)              # Load the lessR package.
help(package=lessR)         # Show the information page.
sessionInfo()               # Confirm all attached packages.

lessR::SummaryStats(Weight, dframe=WeightG12Stu.df)
# Provide a wide variety of summary statistics and identify
# outliers, if any

par(ask=TRUE)
lessR::BoxPlot(Weight, dframe=WeightG12Stu.df,
  main="Boxplot of Weight Generated by the
  lessR::BoxPlot() Function")
# Produce a boxplot and accompanying descriptive statistics
# about the boxplot and add a title to the figure

par(ask=TRUE)
lessR::Histogram(Weight, dframe=WeightG12Stu.df,
  main="Histogram of Weight Generated by the
  lessR::Histogram() Function")
# Produce a histogram and accompanying descriptive statistics
# about the histogram and add a title to the figure

par(ask=TRUE)
lessR::Density(Weight, dframe=WeightG12Stu.df,
  main="Density Curve, Histogram, and Normal Curve of Weight
  Generated by the lessR::Density() Function")
# Produce a density curve, histogram, and normal curve,
# identify accompanying descriptive statistics about the
# density curve, and add a title to the figure
```

2.8 Prepare to Exit, Save, and Later Retrieve This R Session

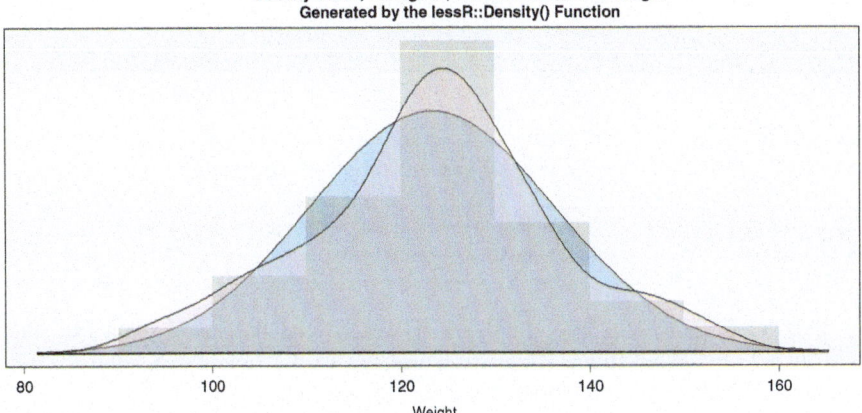

```
install.packages("epicalc")
library(epicalc)            # Load the epicalc package.
help(package=epicalc)       # Show the information page.
sessionInfo()               # Confirm all attached packages.

par(ask=TRUE)
epicalc::summ(WeightG12Stu.df$Weight,
   by=NULL,             # No breakout statistics.
   graph=TRUE,          # Use graph=TRUE, if desired.
   pch=20, ylab="auto",
   main="Sorted Dotplot of Weight Generated by the
   epicalc::summ() Function",
   cex.X.axis=1.25, # Note X axis label size.
   cex.Y.axis=1.25, # Note Y axis label size.
   font.lab=2, dot.col="auto")
# Produce a sorted dotplot and accompanying descriptive
# statistics
```

2.8 Prepare to Exit, Save, and Later Retrieve This R Session

It is common to prepare R syntax in a separate file, using a simple ASCII text editor. If time permits, experiment with Crimson Editor, Tinn-R, or vim, but there are many other possible selections.

Use the following set of actions to exit from the current R session.

```
getwd()                 # Identify the current working directory.
ls()                    # List all objects in the working
                        # directory.
ls.str()                # List all objects, with finite detail.
list.files()            # List files at the PC directory.
```

```
save.image("R_Lesson_DescriptiveStatistics.rdata")

getwd()              # Identify the current working directory.
ls()                 # List all objects in the working
                     # directory.
ls.str()             # List all objects, with finite detail.
list.files()         # List files at the PC directory.

alarm()              # Alarm, notice of upcoming action.
q()                  # Quit this session.
                     # Prepare for Save workspace image? query.
```

Use the R Graphical User Interface (GUI) to load the saved rdata file: **File** and then **Load Workspace**. Otherwise, use the load() function, keying the full pathname, to load the .rdata file and retrieve the session.

Recall, however, that it may be just as useful to simply use a R script file (typically saved as a .txt ASCII-type file) and recreate the analyses and graphics, provided the data files remain available.

Chapter 3
Student's t-Test for Independent Samples

Abstract The purpose of this lesson is to use R to examine differences between groups, specifically by using Student's t-Test for Independent Samples. Overall, Student's t-Test is a very common test for determining differences when a singular measured variable (e.g., IQ score, weight of dairy cow milk production per lactation, length of shark dorsal fin, etc.) is compared to differences between a grouping variable with two breakout groups (e.g., Female v Male humans, Guernsey v Jersey cows, Mako v Great White sharks). The t-Test was developed in the early 1900s by Gosset, as part of quality assurance work for a beverage company, but writing under the pen name *Student*. Student's t-Test is the appropriate test for comparing differences between small samples, typically 30 or fewer, as opposed to samples with more than 30 observations.

Keywords Barplot • Boxplot (box-and-whiskers plot) • Density plot • Dotchart • Histogram • Independent samples • Matched pairs • Student's t-Test • t-statistic • z-statistic

3.1 Background on This Lesson

3.1.1 Description of the Data

This lesson on Student's t-Test for independent samples is taken from a biological survey in a specific, but unnamed, area. The data involve two factor-type variables (Gender and Region) and two variables of interval data, coded as M1 and M2. The exact nature of M1 and M2 are currently unknown and are not needed for this lesson. It is only important to know they are interval data.

Regarding the number of observations (e.g., N) and the use of Student's t-Test, recall that:

- Student's t-Test is still an appropriate and commonly used test with more than 30 observations.
- When N goes beyond 30 observations, t approximates z.

When using Student's t-Test to determine if the difference between two groups is indeed a true difference, or if the difference between the two groups is due only to chance, ideally:

- Both groups should approximate normal distribution.
- It is best if random selection were used for all members of the two groups.

However, Student's t-Test is sufficiently robust such that these two assumptions are often not met, or perhaps not met as rigorously as desired.

Using data from a biological survey, this lesson presents four separate t-Tests, to determine if there are differences in Gender (later recoded to Gender.recode, Female v Male) by M1, Gender by M2, Region (North v South) by M1, and Region by M2. The object variables M1 and M2 are both numeric but their nature (e.g., IQ score, weight of milk production per lactation, length of dorsal fin, etc.) is masked from the individual assigned to this set of analyses. The analysist only knows that the data are of a biological nature and what little information is provided in the Code Book. Further, the dataset is fairly small (N = 30) and there are no missing data.

3.1.2 Null Hypothesis (Ho)

There are four separate Null Hypothesis statements associated with this lesson. Each Null Hypothesis is based a significance level of p <= 0.05.

Null Hypothesis (Ho):

- Ho: There is no statistically significant difference (p <= 0.05) in M1 values by Gender (Female v Male).
- Ho: There is no statistically significant difference (p <= 0.05) in M2 values by Gender (Female v Male).
- Ho: There is no statistically significant difference (p <= 0.05) in M1 values by Region (North v South).
- Ho: There is no statistically significant difference (p <= 0.05) in M2 values by Region (North v South).

Notice how the Null Hypothesis (Ho) uses p <= 0.05. The expression p <= 0.05 is used to identify the declared probability level specific to the Null Hypothesis. That is to say, after calculation of the t-statistic and subsequent calculated p value, with correct interpretation of these statistics there will be a 5% or less probability of an incorrect inference related to differences associated with this test.

3.2 Data Import of a .csv Spreadsheet-Type Data File into R   19

Many exploratory inferential analyses in the biological and social sciences are conducted at p <= 0.05. However, you will see some problems set at the more restrictive p <= 0.01 and even p <= 0.001.

Along with the use of p, you will also see the term alpha in any discussions about the level of probability, but p will be used in this lesson.

Be sure to note how numerical codes have been purposely used for some data (e.g., Gender) in this lesson. Numerical codes are quite common and as such, the creation of a later Code Book is essential so that there is agreement on what each code represents.

3.2 Data Import of a .csv Spreadsheet-Type Data File into R

For this lesson, the dataset has been prepared in .csv (e.g., comma-separated values) file format. The data are separated by commas. The data are not separated by tabs and the data are not separated by spaces.

```
###############################################################
# Housekeeping                          Use for All Analyses
###############################################################
rm(list = ls())      # CAUTION: Remove all files in the working
                     # directory. If this action is not desired,
                     # use the rm() function one-by-one to remove
                     # the objects that are not needed.
setwd("F:/R_Biostatistics")
                     # Set to a new working directory.
                     # Note the single forward slash and double
                     # quotes.
                     # This new directory should be the directory
                     # where the data file is located, otherwise
                     # the data file will not be found.
getwd()              # Confirm the working directory.
search()             # Attached packages and objects.
###############################################################
```

Create an object called GenRegM1M2.df. The object GenRegM1M2.df will be a dataframe, as indicated by the enumerated .df extension to the object name. This object will represent the output of applying the read.table() function against the comma-separated values file called GenderRegionM1M2.csv. Note the arguments used with the read.table() function, showing that there is a header with descriptive variable names (header=TRUE) and that the separator between fields is a comma (sep=",").

```
GenRegM1M2.df <- read.table (file =
  "GenderRegionM1M2.csv",
  header = TRUE,
  sep = ",")                    # Import the .csv file
```

```
getwd()                       # Identify the working directory
ls()                          # List objects
attach(GenRegM1M2.df)         # Attach the data, for later use
str(GenRegM1M2.df)            # Identify structure
nrow(GenRegM1M2.df)           # List the number of rows
ncol(GenRegM1M2.df)           # List the number of columns
dim(GenRegM1M2.df)            # Dimensions of the data frame
names(GenRegM1M2.df)          # Identify names
colnames(GenRegM1M2.df)       # Show column names
rownames(GenRegM1M2.df)       # Show row names
head(GenRegM1M2.df)           # Show the head
tail(GenRegM1M2.df)           # Show the tail
GenRegM1M2.df                 # Show the entire dataframe
summary(GenRegM1M2.df)        # Summary statistics
```

By completing this action, an object called GenRegM1M2.df has been created. This R-based object is a dataframe and it consists of the data originally included in the file GenderRegionM1M2.csv, a comma-separated .csv file. To avoid possible conflicts, make sure that there are no prior R-based objects called GenRegM1M2.df. The prior use of rm(list=ls()) accommodates this concern, removing all prior objects in the current R session.

Note how it was only necessary to key the filename for the .csv file and not the full pathname since the R working directory is currently set to the directory and/or subdirectory where this .csv file is located (see the Housekeeping section at the beginning of this lesson).

3.3 Organize the Data and Display the Code Book

For this lesson, the class() function, str() function, and duplicated() function will be used to be sure that data are organized as desired.

```
class(GenRegM1M2.df)
class(GenRegM1M2.df$Subject)  # DataFrame$ObjectName notation
class(GenRegM1M2.df$Gender)   # DataFrame$ObjectName notation
class(GenRegM1M2.df$Region)   # DataFrame$ObjectName notation
class(GenRegM1M2.df$M1)       # DataFrame$ObjectName notation
class(GenRegM1M2.df$M2)       # DataFrame$ObjectName notation

str(GenRegM1M2.df)            # Structure

duplicated(GenRegM1M2.df)     # Duplicates
```

The class for each object seems to be correct and there are no duplicate rows of data in the dataframe. Saying this, a Code Book will help with future understanding this dataset.

A Code Book is an essential aid for anyone involved in the day-to-day activities of the research and statistics process. The Code Book is typically brief and only

3.3 Organize the Data and Display the Code Book

serves as a useful reminder for what can be easily forgotten months (or even weeks) later, to make it easy to decipher what may otherwise be seen as arcane numeric codes. Coding schemes that are intuitively obvious today can easily become obtuse tomorrow.

The Code Book below represents how data are desired before analyses begin. Recoding may be needed to put data into new formats.

```
########################################################
# Code Book                                             #
########################################################
#                                                      #
# Subject .................... Factor (e.g., nominal) #
#                     A unique ID ranging from 1 to 30 #
#                                                      #
# Gender ......................Factor (e.g., nominal) #
#                                          Female = 1  #
#                                          Male   = 2  #
#                                                      #
# Region ..................... Factor (e.g., nominal) #
#                                                North #
#                                                South #
#                                                      #
# M1 ......................... Numeric (e.g., interval) #
#                   An unidentified biological variable #
#                       that ranges from 0.00 to 100.0 #
#                                                      #
# M2 ......................... Numeric (e.g., interval) #
#                   An unidentified biological variable #
#                      that ranges from 0.000 to 2.000 #
########################################################
```

The str() function is then again applied against the dataframe to see the nature of each object variable as well as confirmation that the data are collectively viewed as a dataframe:

```
str(GenRegM1M2.df)
```

Recall that the Code Book shows data in their desired formats, which often requires some degree of recoding which has not yet occurred.

Once there is agreement that the data were brought into R in correct format, it is usually necessary to organize the data to some degree:

- The object variable Subject is currently viewed as an integer. Some R users may find it best to recode these identification-type numeric values into row names, but in this lesson Subject will instead be recoded into a factor object variable.
- In this lesson note how integer numeric codes (e.g., 1 and 2) have been used in the original file to identify groups for the factor object variable Gender. A set of

simple R-based actions can easily: (1) transform (e.g., recode) the object variable GenRegM1M2.df$Gender into a new object variable, (2) change the recoded object variable from original integer format to enumerated factor format, and (3) apply English text labels for the otherwise cryptic numeric codes (e.g., 1 and 2).
- In contrast, the groups for the factor object variable Region are presented in the original file as standard English text (e.g., North and South).
- Values for M1 are currently whole numbers and as such they are first treated in R as integers. A simple recode action will instead be used to put these values into decimal format.
- Values for M2 are viewed as numeric values.

This transformation (again, typically called a recode action) is needed and the process, using R-based syntax, follows. There may be some unnecessary (perhaps redundant) actions with the following recode activities, but these are purposely done to provide assurance that each variable is in desired format, both original variables and well as the newly-created (e.g., enumerated) variables such as Gender.recode:

```
GenRegM1M2.df$Subject      <- as.factor(
  GenRegM1M2.df$Subject)

GenRegM1M2.df$Gender.recode <- factor(
  GenRegM1M2.df$Gender,
  labels=c("Female", "Male"))
  # Use factor() and not as.factor().

GenRegM1M2.df$Region       <- as.factor(
  GenRegM1M2.df$Region)

GenRegM1M2.df$M1           <- as.numeric(
  GenRegM1M2.df$M1)

GenRegM1M2.df$M2           <- as.numeric(
  GenRegM1M2.df$M2)
getwd()                    # Identify the working directory
ls()                       # List objects
attach(GenRegM1M2.df)      # Attach the data, for later use
str(GenRegM1M2.df)         # Identify structure
nrow(GenRegM1M2.df)        # List the number of rows
ncol(GenRegM1M2.df)        # List the number of columns
dim(GenRegM1M2.df)         # Dimensions of the data frame
names(GenRegM1M2.df)       # Identify names
colnames(GenRegM1M2.df)    # Show column names
rownames(GenRegM1M2.df)    # Show row names
head(GenRegM1M2.df)        # Show the head
tail(GenRegM1M2.df)        # Show the tail
GenRegM1M2.df              # Show the entire dataframe
summary(GenRegM1M2.df)     # Summary statistics
```

3.4 Conduct a Visual Data Check   23

The object variable GenRegM1M2.df$Gender.recode was created by putting the object variable GenRegM1M2.df$Gender into factor format. Labels were then applied in sequential order for this new object, with Female used to represent every occurrence of the numeric value 1 and Male used to represent every occurrence of the numeric value 2.

The object variable GenRegM1M2.df$M1 was recoded from integer format to numeric format by applying the as.numeric() function.

Note the formal nomenclature used in this recode action and the use of DataFrame$ObjectName notation when working with object variables that are part of a dataframe. Note also how the $ symbol is used to separate the name of the dataframe from the name of the Object: DataframeName$ObjectName.

3.4 Conduct a Visual Data Check

Graphics are important for multiple reasons. Throwaway graphics serve as a useful quality assurance tool, identifying data that may be either out-of-range or illogical. Graphics also provide a general sense of outcomes and comparisons between and among variables. Although the precise statistics presented in tables are important to those who regularly work with data, publishable quality graphics are perhaps the most common medium for communication with the general public on research findings.

First, prepare a throwaway graphic of each main variable simply to see general trends and to also serve as a review of the data. Ignore any attempt to overly embellish these initial figures. Recall that these initial graphics are only attempted to see general trends and to also serve as a quality assurance tool.

```
par(ask=TRUE)        # Plot of Subject
plot(GenRegM1M2.df$Subject,
  main="Plot of Subject")

par(ask=TRUE)        # Dotchart of Gender
dotchart(GenRegM1M2.df$Gender,
  main="Dotchart of Gender Before Recode")

par(ask=TRUE)        # Plot of Gender.recode
plot(GenRegM1M2.df$Gender.recode,
  main="Plot of Gender After Recode")

par(ask=TRUE)        # Plot of Region
plot(GenRegM1M2.df$Region,
  main="Plot of Region")
```

```
par(ask=TRUE)        # Dotchart of M1
dotchart(GenRegM1M2.df$M1,
  main="Dotchart of M1")

par(ask=TRUE)        # Histogram of M1
hist(GenRegM1M2.df$M1,
  main="Histogram of M1")

par(ask=TRUE)        # Dotchart of M2
dotchart(GenRegM1M2.df$M2,
  main="Dotchart of M2")

par(ask=TRUE)        # Histogram of M2
hist(GenRegM1M2.df$M2,
  main="Histogram of M2")

par(ask=TRUE)        # Stacked barplot
barplot(table(GenRegM1M2.df$Gender.recode,
  GenRegM1M2.df$Region),
  beside=FALSE,      # Orientation of the barplot
  main="Barplot of Gender (Stacked) and Region")

par(ask=TRUE)        # Side-by-side barplot
barplot(table(GenRegM1M2.df$Gender.recode,
  GenRegM1M2.df$Region),
  beside=TRUE,       # Orientation of the barplot
  main="Barplot of Gender (Side-by-Side) and Region")
```

More complex barplots may help provide details for Gender and Region that are not quite apparent with simple barplots only. And, a legend may help better identify groups and group membership, while still keeping the figures simple. To achieve this aim, use the table() function to create a new object variable that represents a crosstab table of GenRegM1M2.df$Gender.recode by GenRegM1M2.df$Region. Then, note how this newly created object variable (GenderRegionM1M2.crosstab) is used in concert with the barplot() function and the legend argument.

```
GenderRegionM1M2.crosstab <-table(   # Organize the data
GenRegM1M2.df$Gender.recode,         # into a table, to
  GenRegM1M2.df$Region)              # ease later actions.

GenderRegionM1M2.crosstab
str(GenderRegionM1M2.crosstab)        # Object structure
attributes(GenderRegionM1M2.crosstab) # Object attributes

par(ask=TRUE)      # Stacked barplot with legend
barplot(GenderRegionM1M2.crosstab,
```

3.4 Conduct a Visual Data Check

```
  main="Barplot of Gender (Stacked) and Region
  With Legend",
  legend=rownames(GenderRegionM1M2.crosstab),
  beside=FALSE)      # Orientation of the barplot
  # Presentation of the legend needs improvement, but
  # it is now possible to distinguish between groups.

par(ask=TRUE)         # Side-by-side barplot with legend
barplot(GenderRegionM1M2.crosstab,
  main="Barplot of Gender (Side-by-Side) and Region
  With Legend",
  legend=rownames(GenderRegionM1M2.crosstab),
  beside=TRUE)       # Orientation of the barplot
  # Presentation of the legend needs improvement, but
  # it is now possible to distinguish between groups.
```

Following along with the desire for more detail, the hist(), boxplot(), vioplot::vioplot(), and UsingR::simple.violinplot() functions will be used to provide details on the two numeric object variables in this lesson, M1 and M2. Note how information is provided at the singular level of detail and also by breakout groups. Most importantly, note how a few lines of R syntax have been added, providing greater clarity to graphical output while still keeping within the overall desire for fairly easy-to-prepare and reproduce syntax.

```
par(ask=TRUE)
hist(GenRegM1M2.df$M1,
  breaks=10, main="Histogram of M1")

par(ask=TRUE)
boxplot(GenRegM1M2.df$M1,
  horizontal=FALSE,
  main="Vertical Boxplot of M1")

install.packages("vioplot")
library(vioplot)             # Load the vioplot package.
help(package=vioplot)        # Show the information page.
sessionInfo()                # Confirm all attached pac
                             kages.

par(ask=TRUE)
vioplot::vioplot(GenRegM1M2.df$M1,
  horizontal=FALSE, names=c("M1 Values"))
  title("Vertical Violin Plot of M1")

par(ask=TRUE)
boxplot(GenRegM1M2.df$M1 ~    # Note ~ character
  GenRegM1M2.df$Gender.recode,
  main="Boxplot of M1 by Gender")
```

```
par(ask=TRUE)
boxplot(GenRegM1M2.df$M1 ~     # Note ~ character
  GenRegM1M2.df$Region,
  main="Boxplot of M1 by Region")

install.packages("UsingR")
library(UsingR)                # Load the UsingR package.
help(package=UsingR)           # Show the information page.
sessionInfo()                  # Confirm all attached packages.

par(ask=TRUE)
UsingR::simple.violinplot(GenRegM1M2.df$M1 ~
  GenRegM1M2.df$Gender.recode)
  title("Violin Plot of M1 by Gender")

par(ask=TRUE)
UsingR::simple.violinplot(GenRegM1M2.df$M1 ~
  GenRegM1M2.df$Region)
  title("Violin Plot of M1 by Region")

par(ask=TRUE)
hist(GenRegM1M2.df$M2,
  breaks=10, main="Histogram of M2")

par(ask=TRUE)
boxplot(GenRegM1M2.df$M2,
  horizontal=FALSE,
  main="Vertical Boxplot of M2")

par(ask=TRUE)
vioplot::vioplot(GenRegM1M2.df$M2,
  horizontal=FALSE, names=c("M2 Values"))
  title("Vertical Violin Plot of M2")

par(ask=TRUE)
boxplot(GenRegM1M2.df$M2 ~     # Note ~ character
  GenRegM1M2.df$Gender.recode,
  main="Boxplot of M2 by Gender")

par(ask=TRUE)
boxplot(GenRegM1M2.df$M2 ~     # Note ~ character
  GenRegM1M2.df$Region,
  main="Boxplot of M2 by Region")

par(ask=TRUE)
UsingR::simple.violinplot(GenRegM1M2.df$M2 ~
  GenRegM1M2.df$Gender.recode)
  title("Violin Plot of M2 by Gender")
```

3.4 Conduct a Visual Data Check

```
par(ask=TRUE)
UsingR::simple.violinplot(GenRegM1M2.df$M2 ~
  GenRegM1M2.df$Region)
  title("Violin Plot of M2 by Region")
```

After these initial figures are reviewed and when there is agreement that data are correct and the general approach for graphics is acceptable, prepare a more embellished figure if desired. Remember to make colors vibrant and use print that is large and dark, whenever possible, to support future public display of the figure.

There are many R-based functions and arguments to select from when preparing graphics. The base R tools that come with initial download typically meet immediate needs for the production of graphics. However, with practice and more experience be sure to explore the many additional R-based functions and graphics available in the thousands of external packages currently available to the R community. These additional functions and arguments often also produce statistics of some type, to go along with production of the desired figure. Again, explore the many possibilities available here and recall that when using R, rarely if ever is there one-and-only-one function to support production of a desired graphic.

```
par(ask=TRUE)              # Embellished graphic
barplot(table(GenRegM1M2.df$Gender.recode),
  main="Barplot of Gender",                    # Title
  xlab="Gender", ylab="Frequency",             # X and Y axis
  cex.axis=1.25, cex.names=1.25, cex.lab=1.25, # Size
  col="red", font.lab=2)                       # Color - Font

par(ask=TRUE)              # Embellished graphic
barplot(table(GenRegM1M2.df$Region),
  main="Barplot of Region",                    # Title
  xlab="Trait", ylab="Frequency",              # X and Y axis
  cex.axis=1.25, cex.names=1.25, cex.lab=1.25, # Size
  col="red", font.lab=2)                       # Color - Font
```

Although these figures, prepared by using the barplot() function, are both useful and visually appealing, use the epicalc package to produce additional information (e.g., frequency distributions and percentage representation) about the factor object variables in question, Gender.recode and Region for this sample. The additional statistics gained by using these specialized functions are information-rich and give reason why specialized functions, such as epicalc::tableStack() and epicalc::tab1() demand attention.

```
install.packages("epicalc")
library(epicalc)              # Load the epicalc package.
help(package=epicalc)         # Show the information page.
sessionInfo()                 # Confirm all attached packages.

epicalc::tableStack(Gender.recode,
  dataFrame=GenRegM1M2.df,
  by="none", count=TRUE, decimal=2,
  percent=c("column", "row"),
```

```
                  frequency=TRUE, name.test=TRUE,
                  total.column=TRUE, test=TRUE)

par(ask=TRUE)          # Bar Plot of Gender.recode
epicalc::tab1(GenRegM1M2.df$Gender.recode,
   decimal=2,                          # Use the tab1() function
   sort.group=FALSE,                   # from the epicalc
   cum.percent=TRUE,                   # package to see details
   graph=TRUE,                         # about the selected
   missing=TRUE,                       # object variable. (The
   bar.values=c("frequency"),          # 1 of tab1 is the one
   horiz=FALSE,                        # numeric character and
   cex=1.15,                           # it is not the letter
   cex.names=1.15,                     # l).
   cex.lab=1.15, cex.axis=1.15,
   main="Gender:  Breakout N Values",
   ylab="Frequency of Gender, Includings NAs if Any",
   col= c("pink", "blue"),
   gen=TRUE)

epicalc::tableStack(Region,
   dataFrame=GenRegM1M2.df,
   by="none", count=TRUE, decimal=2,
   percent=c("column", "row"),
   frequency=TRUE, name.test=TRUE,
   total.column=TRUE, test=TRUE)

par(ask=TRUE)          # Bar Plot of Region
epicalc::tab1(GenRegM1M2.df$Region,
   decimal=2,                          # Use the tab1() function
   sort.group=FALSE,                   # from the epicalc
   cum.percent=TRUE,                   # package to see details
   graph=TRUE,                         # about the selected
   missing=TRUE,                       # object variable. (The
   bar.values=c("frequency"),          # 1 of tab1 is the one
   horiz=FALSE,                        # numeric character and
   cex=1.15,                           # it is not the letter
   cex.names=1.15,                     # l).
   cex.lab=1.15, cex.axis=1.15,
   main="Region:  Breakout N Values",
   ylab="Frequency of Region, Includings NAs if Any",
   col= c("red", "green"),
   gen=TRUE)
```

With useful information and publishable graphics provided for individual object variables, it is now best to revisit GenderRegionM1M2.crosstab, which was previously created as a crosstabs table of Gender.recode by Region and then use this object throughout, to save time and trouble each time it is needed. However, the barplots will now be presented with appropriate color codes and embellished with vibrant fonts and heavy lines.

3.4 Conduct a Visual Data Check

```
str(GenderRegionM1M2.crosstab)
GenderRegionM1M2.crosstab
summary(GenderRegionM1M2.crosstab)

par(ask=TRUE)      # Barplot of Gender.recode by Region
barplot(GenderRegionM1M2.crosstab,
  main="Gender by Region (Stacked)",          # Title
  xlab="Region", ylab="Frequency",            # X and Y axis
  cex.axis=1.25, cex.names=1.25, cex.lab=1.25,  # Size
  font.lab=2, col=c("pink", "blue"),          # Factor colors
  legend=rownames(GenderRegionM1M2.crosstab), # Legend
  ylim=c(0,20),                               # Y scale
  beside=FALSE)                               # Stacked
  # Note how the argument ylim=c(0,20) was used so that the
  # legend would show in whitespace and not be lost showing
  # against part of a bar.

par(ask=TRUE)      # Barplot of Gender.recode by Region
barplot(GenderRegionM1M2.crosstab,
  main="Gender by Region (Side-by-Side)",     # Title
  xlab="Region", ylab="Frequency",            # X and Y axis
  cex.axis=1.25, cex.names=1.25, cex.lab=1.25,  # Size
  font.lab=2, col=c("red", "green"),          # Factor colors
  legend=rownames(GenderRegionM1M2.crosstab), # Legend
  ylim=c(0,15),                               # Y scale
  beside=TRUE)                                # Side-by-side
  # Note how the argument ylim=c(0,15) was used so that the
  # legend would show in whitespace and not be lost showing
  # against part of a bar.
```

There are also many ways to show the numeric variables, individually and by breakout groups. From among the many possible selections, the lattice package and specifically the lattice::histogram() and lattice::bwplot() functions will be used to show valuable displays of M1 and M2 individually and then by Gender and by Region. A few examples of the many possible combinations are shown below.

```
install.packages("lattice")
library(lattice)              # Load the lattice package.
help(package=lattice)         # Show the information page.
sessionInfo()                 # Confirm all attached packages.

par(ask=TRUE) # 1 Column by 1 Row Histogram
lattice::histogram(~ GenRegM1M2.df$M1,
  type="count", # Note: count
  par.settings=simpleTheme(lwd=2),
  par.strip.text=list(cex=1.15, font=2),
  scales=list(cex=1.15),
  main="Histogram (Count) of M1",
  xlab=list("M1", cex=1.15, font=2),
```

```
  xlim=c(0,120), # Note the range.
  ylab=list("Count", cex=1.15, font=2),
  aspect=1, breaks=10,
  layout = c(1,1), # Note: 1 Column by 1 Row.
  col="red")

par(ask=TRUE) # 1 Column by 2 Rows Histogram
lattice::histogram(~ GenRegM1M2.df$M1 |
  GenRegM1M2.df$Gender.recode,
  type="count", # Note: count
  par.settings=simpleTheme(lwd=2),
  par.strip.text=list(cex=1.15, font=2),
  scales=list(cex=1.15),
  main="Histograms (Count) of M1 by Gender",
  xlab=list("M1", cex=1.15, font=2),
  xlim=c(0,120), # Note the range.
  ylab=list("Count", cex=1.15, font=2),
  aspect=0.25, breaks=10,
  layout = c(1,2), # Note: 1 Column by 2 Rows.
  col="red")

par(ask=TRUE) # 1 Column by 1 Row Histogram
lattice::histogram(~ GenRegM1M2.df$M2,
  type="count", # Note: count
  par.settings=simpleTheme(lwd=2),
  par.strip.text=list(cex=1.15, font=2),
  scales=list(cex=1.15),
  main="Histogram (Count) of M2",
  xlab=list("M2", cex=1.15, font=2),
  xlim=c(0,2.25), # Note the range.
  ylab=list("Count", cex=1.15, font=2),
  aspect=1, breaks=10,
  layout = c(1,1), # Note: 1 Column by 1 Row.
  col="red")

par(ask=TRUE) # 1 Column by 2 Rows Histogram
lattice::histogram(~ GenRegM1M2.df$M2 |
  GenRegM1M2.df$Region,
  type="count", # Note: count
  par.settings=simpleTheme(lwd=2),
  par.strip.text=list(cex=1.15, font=2),
  scales=list(cex=1.15),
  main="Histograms (Count) of M2 by Region",
  xlab=list("M2", cex=1.15, font=2),
  xlim=c(0,2.25), # Note the range.
  ylab=list("Count", cex=1.15, font=2),
  aspect=0.25, breaks=10,
  layout = c(1,2), # Note: 1 Column by 2 Rows.
  col="red")
```

3.4 Conduct a Visual Data Check

```
par(ask=TRUE) # Singular boxplot.
lattice::bwplot(GenRegM1M2.df$M1,
  par.settings = simpleTheme(lwd=2),
  par.strip.text=list(cex=1.15, font=2),
  scales=list(cex=1.15),
  main="Boxplot of M1",
  xlab=list("M1", cex=1.15, font=2),
  xlim=c(0,120), aspect=0.5, layout=c(1,1),
  col="red")

par(ask=TRUE) # Breakout group by measured object
lattice::bwplot(GenRegM1M2.df$Gender.recode ~
  GenRegM1M2.df$M1,
  par.settings = simpleTheme(lwd=2),
  par.strip.text=list(cex=1.15, font=2),
  scales=list(cex=1.15),
  main="Boxplot of M1 by Gender",
  xlab=list("M1", cex=1.15, font=2),
  xlim=c(0,120),
  ylab=list("Gender", cex=1.15, font=2),
  aspect=0.5, layout=c(1,1), col="red")

par(ask=TRUE) # Singular boxplot.
lattice::bwplot(GenRegM1M2.df$M2,
  par.settings = simpleTheme(lwd=2),
  par.strip.text=list(cex=1.15, font=2),
  scales=list(cex=1.15),
  main="Boxplot of M2",
  xlab=list("M2", cex=1.15, font=2),
  xlim=c(0,2.5), aspect=0.5, layout=c(1,1),
  col="red")

par(ask=TRUE) # Breakout group by measured object
lattice::bwplot(GenRegM1M2.df$Region ~
  GenRegM1M2.df$M2,
  par.settings = simpleTheme(lwd=2),
  par.strip.text=list(cex=1.15, font=2),
  scales=list(cex=1.15),
  main="Boxplot of M2 by Region",
  xlab=list("M2", cex=1.15, font=2),
  xlim=c(0,2.5),
  ylab=list("Region", cex=1.15, font=2),
  aspect=0.5, layout=c(1,1), col="red")
```

Although histograms and boxplots are certainly useful graphical tools, the density plot is especially helpful for visualizing how data are distributed individually and by breakout groups.

For this sample, first prepare throwaway density plots with the UsingR::Density Plot() function and then, if the outcomes show promise, use the lattice::densityplot() function for more aesthetic and visually appealing density plot images.

```
par(ask=TRUE)
UsingR::DensityPlot(M1 ~ Gender.recode, data=GenRegM1M2.df)

par(ask=TRUE)
UsingR::DensityPlot(M1 ~ Region, data=GenRegM1M2.df)

par(ask=TRUE)
UsingR::DensityPlot(M2 ~ Gender.recode, data=GenRegM1M2.df)

par(ask=TRUE)
UsingR::DensityPlot(M2 ~ Region, data=GenRegM1M2.df)
```

As demonstrated above, a density plot is certainly a useful tool to visualize how measured data are organized, at the singular level and also by breakout groups. Below, look at the way these density plots have been prepared again, but now with more detail and presentation by using the lattice::densityplot() function. Although statistical analyses are certainly necessary and should never be neglected, no matter how visual presentations appear, in this lesson note how the density plot presentation of M2 by Region brings to attention possible differences. Significant differences may also be in play for other comparisons, but the visualization of M2 by Region is striking and gives an early view of areas that need further attention.

```
par(ask=TRUE) # 1 Column by 1 Row Density Plot
lattice::densityplot(~ GenRegM1M2.df$M1,
  type="count", # Note: count
  par.settings=simpleTheme(lwd=2),
  par.strip.text=list(cex=1.15, font=2),
  scales=list(cex=1.15),
  main="Density Plot of M1",
  xlab=list("M1", cex=1.15, font=2),
  xlim=c(0,120), # Note the range.
  ylab=list("Density", cex=1.15, font=2),
  aspect=1,
  layout = c(1,1), # Note: 1 Column by 1 Row.
  col="red")

par(ask=TRUE) # 1 Column by 2 Rows Density Plot
lattice::densityplot(~ GenRegM1M2.df$M1 |
  GenRegM1M2.df$Gender.recode,
  type="count", # Note: count
  par.settings=simpleTheme(lwd=2),
  par.strip.text=list(cex=1.15, font=2),
  scales=list(cex=1.15),
```

3.4 Conduct a Visual Data Check

```
  main="Density Plot of M1 by Gender",
  xlab=list("M1", cex=1.15, font=2),
  xlim=c(0,120), # Note the range.
  ylab=list("Density", cex=1.15, font=2),
  aspect=0.25,
  layout = c(1,2), # Note: 1 Column by 2 Rows.
  col="red")

par(ask=TRUE) # 1 Column by 2 Rows Density Plot
lattice::densityplot(~ GenRegM1M2.df$M1 |
  GenRegM1M2.df$Region,
  type="count", # Note: count
  par.settings=simpleTheme(lwd=2),
  par.strip.text=list(cex=1.15, font=2),
  scales=list(cex=1.15),
  main="Density Plot of M1 by Region",
  xlab=list("M1", cex=1.15, font=2),
  xlim=c(0,120), # Note the range.
  ylab=list("Density", cex=1.15, font=2),
  aspect=0.25,
  layout = c(1,2), # Note: 1 Column by 2 Rows.
  col="red")

par(ask=TRUE) # 1 Column by 1 Row Density Plot
lattice::densityplot(~ GenRegM1M2.df$M2,
  type="count", # Note: count
  par.settings=simpleTheme(lwd=2),
  par.strip.text=list(cex=1.15, font=2),
  scales=list(cex=1.15),
  main="Density Plot of M2",
  xlab=list("M2", cex=1.15, font=2),
  xlim=c(0,2.25), # Note the range.
  ylab=list("Density", cex=1.15, font=2),
  aspect=1,
  layout = c(1,1), # Note: 1 Column by 1 Row.
  col="red")

par(ask=TRUE) # 1 Column by 2 Rows Density Plot
lattice::densityplot(~ GenRegM1M2.df$M2 |
  GenRegM1M2.df$Gender.recode,
  type="count", # Note: count
  par.settings=simpleTheme(lwd=2),
  par.strip.text=list(cex=1.15, font=2),
  scales=list(cex=1.15),
  main="Density Plot of M2 by Gender",
```

```
     xlab=list("M2", cex=1.15, font=2),
     xlim=c(0,2.25), # Note the range.
     ylab=list("Density", cex=1.15, font=2),
     aspect=0.25,
     layout = c(1,2), # Note: 1 Column by 2 Rows.
     col="red")

par(ask=TRUE) # 1 Column by 2 Rows Density Plot
lattice::densityplot(~ GenRegM1M2.df$M2 |
  GenRegM1M2.df$Region,
  type="count", # Note: count
  par.settings=simpleTheme(lwd=2),
  par.strip.text=list(cex=1.15, font=2),
  scales=list(cex=1.15),
  main="Density Plot of M2 by Region",
  xlab=list("M2", cex=1.15, font=2),
  xlim=c(0,2.25), # Note the range.
  ylab=list("Density", cex=1.15, font=2),
  aspect=0.25,
  layout = c(1,2), # Note: 1 Column by 2 Rows.
  col="red")
```

Of course, the lattice package is by no means the only package with functions that support breakout group comparisons for density plots. As shown previously, consider the UsingR::DensityPlot() function as a first tool for preparation of throwaway density-type graphical comparisons.

Although descriptive statistics and inferential tests (Student's t-Test for this lesson) are needed to make final judgment, the figures provide a fairly good idea of general trends and how the data compare to each other, individually and by group breakouts. Remember that the syntax used in this lesson can of course be used with future analyses. Simply alter the syntax, typically the dataframe name and object names, and adjust margins as needed to account for scales.

3.5 Descriptive Analysis of the Data

Given the different ways missing data can impact analyses, it is often helpful to first check for missing data (which are not present in this lesson) by using the is.na() function and the complete.cases() function against the entire dataset. Both functions return a TRUE or FALSE response, depending on the function and the outcome of whether data are missing or data are not missing.

```
is.na(GenRegM1M2.df)            # Check for missing data
complete.cases(GenRegM1M2.df)   # Check for complete cases
```

3.5 Descriptive Analysis of the Data

For this simple dataset, the summary() function may be all that is necessary to gain a sense of the data. Note how the summary() function is applied against the entire dataset, thus yielding information about all object variables including those that are not directly used in this sample, including ostensibly unnecessary information about Subject and also Gender, prior to Gender's recode into Gender.recode.

```
summary(GenRegM1M2.df)
```

Although the summary() function is quite sufficient, descriptive statistics for individual object variables may be desired. To achieve this aim, review the prior lesson *Descriptive Statistics and Measures of Central Tendency* for a comprehensive review of the functions used for descriptive statistics, especially: length(), asbio::Mode(), median(), mean(), sd(), table(), and finally summary().

```
length(GenRegM1M2.df$Gender.recode)     # N
length(GenRegM1M2.df$Region)            # N
length(GenRegM1M2.df$M1)                # N
length(GenRegM1M2.df$M2)                # N

install.packages("asbio")
library(asbio)               # Load the asbio package.
help(package=asbio)          # Show the information page.
sessionInfo()                # Confirm all attached packages.

asbio::Mode(GenRegM1M2.df$M1)           # Mode
asbio::Mode(GenRegM1M2.df$M2)           # Mode

median(GenRegM1M2.df$M1, na.rm=TRUE)    # Median
median(GenRegM1M2.df$M2, na.rm=TRUE)    # Median

mean(GenRegM1M2.df$M1, na.rm=TRUE)      # Mean
sd(GenRegM1M2.df$M1,na.rm=TRUE )        # SD
  # Measures of Central Tendency

mean(GenRegM1M2.df$M2,na.rm=TRUE )      # Mean
sd(GenRegM1M2.df$M2, na.rm=TRUE)        # SD
  # Measures of Central Tendency

table(GenRegM1M2.df$Gender.recode)
  # Frequency Distribution of Nominal Variable
table(GenRegM1M2.df$Region)
  # Frequency Distribution of Nominal Variable

summary(GenRegM1M2.df)
```

The epicalc::summ() function is also useful in that it can provide descriptive statistics and a representative figure of individual object variables.

```
par(ask=TRUE) # Use the epicalc package.
epicalc::summ(GenRegM1M2.df$M1,
```

```r
    by=NULL, graph=TRUE, box=TRUE,    # Make a boxplot
    pch=18, ylab="auto",
    main="Sorted Dotplot and Boxplot of M1",
    cex.X.axis=1.15, cex.Y.axis=1.15, font.lab=2,
    dot.col="auto")
    # Note the descriptive statistics that go
    # along with the  epicalc::summ() function.

par(ask=TRUE) # Use the epicalc package.
epicalc::summ(GenRegM1M2.df$M1,
    by=GenRegM1M2.df$Gender.recode,
    graph=TRUE, box=FALSE,            # No boxplot
    pch=18, ylab="auto",
    main="Sorted Dotplot M1 by Gender",
    cex.X.axis=1.15, cex.Y.axis=1.15, font.lab=2,
    dot.col="auto")
    # Note the descriptive statistics that go
    # along with the epicalc::summ() function.

par(ask=TRUE) # Use the epicalc package.
epicalc::summ(GenRegM1M2.df$M1,
    by=GenRegM1M2.df$Region,
    graph=TRUE, box=FALSE,            # No boxplot
    pch=18, ylab="auto",
    main="Sorted Dotplot M1 by Region",
    cex.X.axis=1.15, cex.Y.axis=1.15, font.lab=2,
    dot.col="auto")
    # Note the descriptive statistics that go
    # along with the epicalc::summ() function.

par(ask=TRUE) # Use the epicalc package.
epicalc::summ(GenRegM1M2.df$M2,
    by=NULL, graph=TRUE, box=TRUE,    # Make a boxplot
    pch=18, ylab="auto",
    main="Sorted Dotplot and Boxplot of M2",
    cex.X.axis=1.15, cex.Y.axis=1.15, font.lab=2,
    dot.col="auto")
    # Note the descriptive statistics that go
    # along with the  epicalc::summ() function.

par(ask=TRUE) # Use the epicalc package.
epicalc::summ(GenRegM1M2.df$M2,
    by=GenRegM1M2.df$Gender.recode,
    graph=TRUE, box=FALSE,            # No boxplot
    pch=18, ylab="auto",
```

3.5 Descriptive Analysis of the Data

```
  main="Sorted Dotplot M2 by Gender",
  cex.X.axis=1.15, cex.Y.axis=1.15, font.lab=2,
  dot.col="auto")
  # Note the descriptive statistics that go
  # along with the epicalc::summ() function.

par(ask=TRUE) # Use the epicalc package.
epicalc::summ(GenRegM1M2.df$M2,
  by=GenRegM1M2.df$Region,
  graph=TRUE, box=FALSE,              # No boxplot
  pch=18, ylab="auto",
  main="Sorted Dotplot M2 by Region",
  cex.X.axis=1.15, cex.Y.axis=1.15, font.lab=2,
  dot.col="auto")
  # Note the descriptive statistics that go
  # along with the epicalc::summ() function.
```

The epicalc::summ() by=NULL argument can be set to either TRUE or FALSE, to either obtain or limit breakout descriptive statistics, as desired.

Although the epicalc::summ() function may be sufficient for production of descriptive statistics by different groups, there are many other functions that serve the same purpose. A few will be demonstrated, including the tapply() function and the psych::describeBy() function. As time permits, explore the many other R functions that serve a similar purpose.

```
tapply(M1, Gender.recode, summary, na.rm=TRUE,
  data=GenRegM1M2.df) # Breakouts of M1 by Gender.recode

tapply(M2, Gender.recode, summary, na.rm=TRUE,
  data=GenRegM1M2.df) # Breakouts of M2 by Gender.recode

tapply(M1, Region, summary, na.rm=TRUE,
  data=GenRegM1M2.df) # Breakouts of M1 by Region

tapply(M2, Region, summary, na.rm=TRUE,
  data=GenRegM1M2.df) # Breakouts of M2 by Region

install.packages("psych")
library(psych)             # Load the psych package.
help(package=psych)        # Show the information page.
sessionInfo()              # Confirm all attached packages.

psych::describeBy(GenRegM1M2.df$M1,
  GenRegM1M2.df$Gender.recode)
  # Breakouts of M1 by Gender.recode

psych::describeBy(GenRegM1M2.df$M2,
  GenRegM1M2.df$Gender.recode)
  # Breakouts of M2 by Gender.recode
```

```
psych::describeBy(GenRegM1M2.df$M1,
  GenRegM1M2.df$Region)
# Breakouts of M1 by Region

psych::describeBy(GenRegM1M2.df$M2,
  GenRegM1M2.df$Region)
# Breakouts of M2 by Region
```

The tables::tabular() function can be used to provide even more detail, in a fairly attractive table format that can be easily copied or used in some other fashion in a summary report.

```
install.packages("tables")
library(tables)            # Load the tables package.
help(package=tables)       # Show the information page.
sessionInfo()              # Confirm all attached packages.

tables::tabular((Gender.recode + 1) ~ (n=1) + Format(digits=2)*
  (M1 + M2)*(min + max + mean + sd),
  data=GenRegM1M2.df)
# Gender.recode (row) by M1 and M2 (columns)
```

Note how simple it would be to merely copy and then paste the output from R to an editor, presentation program, etc.

		M1				M2			
Gender.recode	n	min	max	mean	sd	min	max	mean	sd
Female	15	45.00	95.00	79.67	14.13	1.09	1.81	1.58	0.24
Male	15	57.00	97.00	86.80	9.75	1.13	1.81	1.46	0.24
All	30	45.00	97.00	83.23	12.47	1.09	1.81	1.52	0.24

Then, reuse the existing R-based syntax but now substitute as needed to focus on output by Region instead of output by Gender.recode.

```
tables::tabular((Region + 1) ~ (n=1) + Format(digits=2)*
  (M1 + M2)*(min + max + mean + sd),
  data=GenRegM1M2.df)
# Region (row) by M1 and M2 (columns)
```

Again, the output from the tables::tabular() function is quite useful and easily moved from R to an editor or presentation program by using simple copy and paste actions.

		M1				M2			
Region	n	min	max	mean	sd	min	max	mean	sd
North	14	45.000	97.000	81.786	15.572	1.650	1.810	1.744	0.046
South	16	56.000	96.000	84.500	9.309	1.090	1.630	1.333	0.169
All	30	45.000	97.000	83.233	12.467	1.090	1.810	1.525	0.243

For those with specialized publishing purposes, the table output can also be passed through the Hmisc::latex() function to generate output suitable for the LaTeX word processing environment.

3.5 Descriptive Analysis of the Data

```
install.packages("Hmisc")
library(Hmisc)                 # Load the Hmisc package.
help(package=Hmisc)            # Show the information page.
sessionInfo()                  # Confirm all attached packages.

Hmisc::latex(
tables::tabular(((Gender.recode + 1) ~ (n=1) + Format(digits=2)*
   (M1 + M2)*(min + max + mean + sd),
   data=GenRegM1M2.df)
   # Gender.recode (row) by M1 and M2 (columns)
   # in LaTeX-ready format
)

Hmisc::latex(
tables::tabular(((Region + 1) ~ (n=1) + Format(digits=2)*
   (M1 + M2)*(min + max + mean + sd),
   data=GenRegM1M2.df)
   # Region (row) by M1 and M2 (columns) in LaTeX-ready format
) # Wrap LaTeX around output, for those who use LaTex
```

From among the many possible R functions used for descriptive statistics, whether statistics are presented as singular values or in some type of table format, consider the prettyR::brkdn() function also.

```
install.packages("prettyR")
library(prettyR)               # Load the prettyR package.
help(package=prettyR)          # Show the information page.
sessionInfo()                  # Confirm all attached packages.

prettyR::brkdn(M1 ~ Gender.recode,
   data=GenRegM1M2.df, maxlevels=2,
   num.desc=c("mean", "sd", "valid.n"),
   width=25, round.n=2)
   # Gender.recode (row) by M1 (column)

prettyR::brkdn(M2 ~ Gender.recode,
   data=GenRegM1M2.df, maxlevels=2,
   num.desc=c("mean", "sd", "valid.n"),
   width=25, round.n=2)
   # Gender.recode (row) by M2 (column)

prettyR::brkdn(M1 ~ Region,
   data=GenRegM1M2.df, maxlevels=2,
   num.desc=c("mean", "sd", "valid.n"),
   width=25, round.n=2)
   # Region (row) by M1 (column)

prettyR::brkdn(M2 ~ Region,
   data=GenRegM1M2.df, maxlevels=2,
   num.desc=c("mean", "sd", "valid.n"),
```

```
width=25, round.n=2)
# Region (row) by M2 (column)
```

Additional functions could be demonstrated, but the above functions should provide a broad representation of how descriptive statistics and measures of central tendency are determined when using R. With sufficient exposure, individual choice and need determines which functions to use.

3.6 Conduct the Statistical Analysis

Be sure to notice that the dataset for this lesson was prepared in stacked format only. As such, analyses for this lesson will not be approached from the perspective of unstacked data. As a sidebar comment, review the help pages for the stack() function and the unstack() function to learn more about this issue, which is not a concern for this specific dataset but may be a concern for future analyses.

To provide some limit to the length of this lesson, assume that the focus of this demonstration of Student's t-Test is on analysis of differences in M1 by Gender.recode, M2 by Gender.recode, M1 by Region, and M2 by Region. This dataset would certainly support other analyses, but these other analyses are beyond the scope of this lesson.

Note: Use the recoded object variable Gender.recode, to have output show as English text instead of a cryptic letter or numeric code, which would be the case if the object variable Gender, alone, were used. Further, see the assumption below on the nature of equal variance. Consult the prior figures to gain a better perspective.

....

Ho: There is no statistically significant difference ($p <= 0.05$) in M1 values by Gender (Female v Male).

```
t.test(GenRegM1M2.df$M1 ~              # Measured variable
  GenRegM1M2.df$Gender.recode,         # Grouping variable
  alternative="two.sided",             # Two-sided t-Test
  paired=FALSE,                        # Independent
  na.rm=TRUE,                          # Missing data
  var.equal=TRUE)                      # Equal variance

Outcome:   t = -1.6094, df = 28, p-value = 0.1188
mean in group Female    mean in group Male
            79.66667                86.80000
```

Calculated p-value=0.1188, which is greater than $p <= 0.05$. There is no statistically significant difference in M1 values by Gender (Female v Male).

....
....

3.6 Conduct the Statistical Analysis

Ho: There is no statistically significant difference (p <= 0.05) in M2 values by Gender (Female v Male).

```
t.test(GenRegM1M2.df$M2 ~          # Measured variable
  GenRegM1M2.df$Gender.recode,     # Grouping variable
  alternative="two.sided",         # Two-sided t-Test
  paired=FALSE,                    # Independent
  na.rm=TRUE,                      # Missing data
  var.equal=TRUE)                  # Equal variance
```

Outcome: t = 1.3728, df = 28, p-value = 0.1807
mean in group Female   mean in group Male
         1.584667   1.464667

Calculated p-value=0.1807, which is greater than p <= 0.05. There is no statistically significant difference in M2 values by Gender (Female v Male).

....
....

Ho: There is no statistically significant difference (p <= 0.05) in M1 values by Region (North v South).

```
t.test(GenRegM1M2.df$M1 ~          # Measured variable
  GenRegM1M2.df$Region,            # Grouping variable
  alternative="two.sided",         # Two-sided t-Test
  paired=FALSE,                    # Independent
  na.rm=TRUE,                      # Missing data
  var.equal=TRUE)                  # Equal variance
```

Outcome: t = -0.5882, df = 28, p-value = 0.5611
mean in group North mean in group South
         81.78571   84.50000

Calculated p-value=0.5611, which is greater than p <= 0.05. There is no statistically significant difference in M1 values by Region (North v South).

....
....

Ho: There is no statistically significant difference (p <= 0.05) in M2 values by Region (North v South).

```
t.test(GenRegM1M2.df$M2 ~          # Measured variable
  GenRegM1M2.df$Region,            # Grouping variable
  alternative="two.sided",         # Two-sided t-Test
  paired=FALSE,                    # Independent
  na.rm=TRUE,                      # Missing data
  var.equal=TRUE)                  # Equal variance
```

```
Outcome:   t = 8.8014, df = 28, p-value = 1.49e-09
mean in group North mean in group South
          1.743571                1.333125
```

Calculated p-value = 1.49e − 09 (note the e-notation), which is less than p <= 0.05. There is a statistically significant difference in M2 values by Region (North v South).

....

Give notice to the means and p-values for each of these four separate t-Test analyses. Attention to the p-value is perhaps the easiest way to view differences for these analyses.

3.7 Summary

In this lesson, the graphics and statistics provided a great deal of information. Of immediate importance, however, focus on the four Null Hypothesis statements and the outcomes of each.

Although more detail about the nature of M1 and M2 would have been useful, for the immediate purpose of this lesson this detail is not needed and may only cause confusion, which is why the analysist knew very little about the biological survey. What is important to know is that Gender and subsequently Gender.recode as well as Region are factor-type object variables, each with two breakout values. Equally, it is important to know that M1 and M2 are both numeric object variables.

Saying this, the analyses supported the observation that there is no statistically significant difference (p <= 0.05) in M1 values by Gender, M2 values by Gender, and M1 values by Region.

There is, however, a statistically significant difference (p <= 0.05) in M2 values by Region. Along with the statistical output generated from the Student's t-Test, notice the many figures where M2 values are presented by Region to gain a better sense of the differences in M2 values by Region.

It may be redundant to end this lesson with syntax for a figure that was previously provided, but look again at use of the lattice::densityplot() function and how the graphical output provides a sense (and only a sense) of future Null Hypothesis outcomes, as determined by use of Student's t-Test. This figure will focus on the one case where a statistically significant difference was observed, or M2 by Region in this lesson.

Ho: There is no statistically significant difference (p <= 0.05) in M2 values by Region (North v South).

```
par(ask=TRUE) # 1 Column by 2 Rows Density Plot
lattice::densityplot(~ GenRegM1M2.df$M2 |
  GenRegM1M2.df$Region,
  type="count", # Note: count
  par.settings=simpleTheme(lwd=4),
```

```
par.strip.text=list(cex=1.15, font=2),
scales=list(cex=1.15),
main="Density Plot of M2 by Region",
xlab=list("M2 by Region:  t = 8.8014, df = 28,
p-value = 1.49e-09",
cex=1.15, font=2),
xlim=c(0,2.25), # Note the range.
ylab=list("Density", cex=1.15, font=2),
aspect=0.25,
layout = c(1,2), # Note: 1 Column by 2 Rows.
col="red")
```

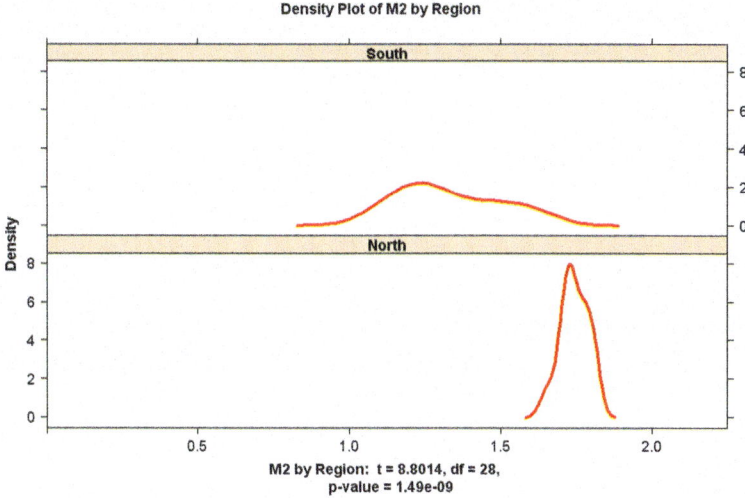

Graphical output and statistical output should always be used in tandem, regardless of future reporting requirements and required form and style. For scientific report writing, the statistical outcomes are typically of primary importance in terms of what is included in a publication. Yet, for presentation purposes, graphical images are desired. The ideal analysis includes both and is then tailored to meet specific presentation needs, whether print or visual.

3.8 Addendum: t-Statistic v z-Statistic

It was previously mentioned that the t-Statistic (e.g., t) begins to approximate the z-Statistic (e.g., z) when the number of subjects in a sample increases, especially after 30 subjects. Look at the enumerated dataset below and determine if the data (Systolic Blood Pressure by Gender) support this statement.

3.8.1 Create the Enumerated Dataset

Use R-based syntax to create a hypothetical dataset, sufficient to meet the needs of this demonstration. Because sampling is used in this demonstration, the composition of the dataset will change each time this syntax is used. Correspondingly, results for the t-Test and z-Test will also change each time.

```
Gender <- c(rep("Male", 1200), rep("Female", 800))
str(Gender)
length(Gender)
Gender
table(Gender)

Gender <- sample(Gender, 2000)
str(Gender)
length(Gender)
Gender
table(Gender)

Systolic <- round(rnorm(2000, 120, 25))
str(Systolic)
length(Systolic)
Systolic
summary(Systolic)

BloodPressure.df <- data.frame(Gender, Systolic)

attach(BloodPressure.df)
str(BloodPressure.df)
BloodPressure.df
summary(BloodPressure.df)
```

3.8.2 Calculate the t-Statistic

Use R syntax to conduct a Student's t-Test of variables in the dataset that was just created. Sampling was used to create this sample dataset so remember that values change each time the dataset in this addendum is created.

```
t.test(BloodPressure.df$Systolic ~       # Measured variable
   BloodPressure.df$Gender,              # Grouping variable
   alternative="two.sided",              # Two-sided t-Test
   paired=FALSE,                         # Independent
   na.rm=TRUE,                           # Missing data
   var.equal=TRUE)                       # Equal variance
```

```
Outcome:   t = 0.578, df = 1998, p-value = 0.5633
mean in group Female    mean in group Male
            120.0037                119.3425
```

3.8.3 Calculate the z-Statistic

Use the coin package and specifically the coin::independence_test() function to conduct a z-Test against variables in the dataset that was just created.

```
install.packages("coin")
library(coin)           # Load the coin package.
help(package=coin)      # Show the information page.
sessionInfo()           # Confirm all attached packages.

coin::independence_test(Systolic ~ Gender,
  data=BloodPressure.df)

Outcome:   Systolic by Gender (Female, Male)
Z = 0.5781, p-value = 0.5632
```

Notice, in this sample of 2,000 subjects, that within the scope of rounding, t approximates z and the two p-values are also approximately equal:

```
t-statistic = 0.578   and p-value = 0.5633
z-statistic = 0.5781 and p-value = 0.5632
```

Caution: Again, results are dependent on composition of the dataset, which was created by using the R-based sample() function. The t-Statistic and z-Statistic change as the dataset changes but it is expected that t and z will approximate each other each time the syntax in this addendum is used.

Student's t-Test is certainly appropriate for small samples (N <= 30 subjects), but the z-statistic may be of more interest when the number of subjects is fairly large.

3.9 Prepare to Exit, Save, and Later Retrieve This R Session

It is common to prepare R syntax in a separate file, using a simple ASCII text editor. If time permits, experiment with Crimson Editor, Tinn-R, or vim, but there are many other possible selections.

Use the following set of actions to exit from the current R session.

```
getwd()            # Identify the current working directory.
ls()               # List all objects in the working
                   # directory.
```

```
ls.str()              # List all objects, with finite detail.
list.files()          # List files at the PC directory.

save.image("R_Lesson_t-Test_Independent-Samples.rdata")

getwd()               # Identify the current working directory.
ls()                  # List all objects in the working
                      # directory.
ls.str()              # List all objects, with finite detail.
list.files()          # List files at the PC directory.

alarm()               # Alarm, notice of upcoming action.
q()                   # Quit this session.
                      # Prepare for Save workspace image? query.
```

Use the R Graphical User Interface (GUI) to load the saved rdata file: **File** and then **Load Workspace**. Otherwise, use the load() function, keying the full pathname, to load the .rdata file and retrieve the session.

Recall, however, that it may be just as useful to simply use a R script file (typically saved as a .txt ASCII-type file) and recreate the analyses and graphics, provided the data files remain available.

Chapter 4
Student's t-Test for Matched Pairs

Abstract The purpose of this lesson is to use R to examine differences to a singular measured variable between pairs, specifically by using Student's t-Test for Matched Pairs. Along with the use of Student's t-Test to compare differences between two separate groups against a singular measured variable, Student's t-Test can also be used to compare differences to a single measured variable when subjects are matched against a counterpart. This lesson also provides an introduction to the use of unstacked data as compared to the use of stacked data. Finally, the issues of sample size (especially N of 30 or so) and sample representation are introduced in this lesson.

Keywords Independent samples • Matched pairs • Pre-test • Post-test • Repeated measures • Stacked data • Student's t-Test • t-statistic • Unstacked data

4.1 Background on This Lesson

4.1.1 Description of the Data

This lesson on Student's t-Test for Matched Pairs has been designed to determine if there is a statistically significant difference (p <= 0.05) in the weight (Kg) of individual biological specimens after the application of a mineral supplement in the feeding program. The individual charged with responsibility for data analysis knows very little about the specifics of the data, the nature of the biological specimen, the feeding program, and the added supplement.

For now, it is only sufficient to know that the biological specimens are all adults of both genders and that they are on a maintenance feeding program. That is to say, they receive sufficient feed and water to maintain their weight, but an overt attempt at weight gain is not a goal for the current feeding program. The feeding program, access to water, available rations, ventilation and housing, etc. remained constant and consistent throughout the experimental period.

Individual specimens were all weighed on the same day. Immediately after weighing the specimens a mineral supplement was introduced into the feeding program. The mineral supplement is tasteless and odorless, has no noticeable texture, and it is assumed to have no impact on feed palatability. After a set, but unidentified, period of time, individual subjects are weighed again, all on the same day. Weights are in Kg and this measurement is viewed as a numeric value. Note how the specimens are not divided into breakout groups (e.g., there is no grouping by gender, breed, color, height at the withers, etc.). Equally, the feeding program and application of the mineral supplement are also consistent, with no grouping or special treatment of any type.

This arrangement is a representation of a simple pre-test and post-test experiment and by design, Student's t-Test for Matched Pairs will be used for statistical analysis:

- Subjects are organized into one common group where individual subjects are matched against their initial pre-test measurement and their final post-test measurement.
- Subjects are measured with an initial pre-test. In this case, the pre-test was weight (Kg) at beginning of the treatment (e.g., feeding program).
- Subjects then experience some type of treatment. In this case, the treatment was an added mineral supplement.
- The nature of the treatment and duration of the treatment is the same for all subjects.
- Subjects are measured with a terminal post-test. In this case, the post-test was weight (Kg) at the end of the treatment (e.g., feeding program).

The matched pairs for this experiment are the pre-test and post-test weights for each individual subject. That is to say, individual subjects are their own matched pair, with weight KgPreSupplement taken at the beginning of the feeding program and weight KgPostSupplement taken again at the end of the treatment.

Review prior discussion on the development, nature, and use of Student's t-Test. The pre-test and post-test approach (e.g., repeated measures) used in this lesson is a common approach for Student's t-Test for Matched Pairs. As a sidebar comment, closely related subjects are often used as matched pairs, such as classmates who experienced similar instruction, human siblings, livestock litter mates, etc., but this is not the approach used in this lesson. In this lesson, individual subjects are their own match, in a pre-test and post-test situation.

This sample will be interesting in that it is a fairly large dataset of 150 subjects and it also introduces the realities of working with missing data. There is a missing post-test datum for Subject 19. There is always a host of reasons why data may be missing from a dataset: the subject may have been unavailable for measurement, the subject may have died, the field technician may have recorded a weight that is simply not valid (e.g., illogical or out-of-range), field notes for this measurement

4.1 Background on This Lesson

may have been either lost or damaged beyond use, the data entry specialist may have failed to enter the datum, etc. Missing data are never desirable, but it is a condition that must be considered since it is common.

Given the nature of the data for this lesson, it is judged that a Student's t-Test for Matched Pairs is an appropriate approach in the attempt to determine if there is any statistically significant difference in weights (Kg) between pre-test measures and post-test measures: KgPreSupplement and KgPostSupplement. Again, recall that there are no planned breakout analyses by gender, varying supplement strengths, varying supplement application periods, etc.

4.1.2 Null Hypothesis (Ho)

There is no statistically significant difference ($p <= 0.05$) in the pre-test weight (Kg) of unidentified adult biological specimens and the post-test weight (Kg) of unidentified adult biological specimens at the termination of a feeding program, where a tasteless and odorless mineral supplement was introduced into the feeding program.

4.1.3 Unstacked Data and Stacked Data

Data for Student's t-Test typically involves identification of the grouping variable (e.g., binary variable; Male v Female, Location 1 v Location 2, supplemental light in the housing unit v natural light only in the housing unit, etc.) and the measured variable. There are two ways that data can be presented in electronic format for a Student's t-Test analysis and then used in R. Again, data can be unstacked and data can be stacked. An example about dogs, which is different from the focus of this lesson, will help clarify these two methods for data organization.

Unstacked Data

Unstacked data for ten separate subjects (2-year old Labrador Retrievers) for gender and weight (Lbs.) follow:

```
Weight.LabDog.Female        Weight.LabDog.Male
51                          55
53                          61
50                          58
49                          60
52                          64
```

With unstacked data, the two columns are typically next to each other. The standard way to use R to conduct a Student's t-Test with unstacked data is to use a comma to separate identification of the two data columns:

```
t.test(Column1, Column2)     # Unstacked data uses , and not ~
t.test(Weight.LabDog.Female, Weight.LabDog.Male)
```

Of course, arguments can always be added to this general means of conducting a Student's t-Test.

Stacked Data

Although the experienced researcher will be able to work with data organized in either format (unstacked or stacked), it is perhaps more common to see data put into stacked format than unstacked format.

Stacked data for ten separate subjects (2-year old Labrador Retrievers) for gender and weight (Lbs.) follow:

```
Gender                 Weight.LabDog
Female                 51
Female                 53
Female                 50
Female                 49
Female                 52
Male                   55
Male                   61
Male                   58
Male                   60
Male                   64
```

The grouping variable, also called the binary variable, is Gender (e.g., Female and Male). The measured datum is Weight.LabDog, measured in Lbs.

With stacked data, as shown above, the data are organized so that data for the grouping variable (Gender, Female and Male) are placed in one column and the corresponding measured data (Weight.LabDog, measured in Lbs.) are found in an adjacent column.

The standard way to use R to conduct a Student's t-Test with stacked data is to use a tilde to separate the variable representing the measured data and the variable representing the two groups.

```
t.test(Measured.Data.Variable ~ Grouping.Variable)
t.test(Weight.LabDog ~ Gender)    # Stacked data uses ~ and not ,
```

Of course, arguments can always be added to this general means of conducting a Student's t-Test.

4.2 Data Import of a .csv Spreadsheet-Type Data File into R

For this lesson, the data have been organized as unstacked data. The dataset has been prepared in .csv (e.g., comma-separated values) file format. The data are separated by commas. The data are not separated by tabs and the data are not separated by spaces.

Assume that the data were originally transferred from paper field notes into a Gnumeric-based spreadsheet, saved in .gnumeric file format. After the spreadsheet was reviewed by staff and put into final form, it was then saved in .csv file format, to facilitate data import action into R and transfer in a common format to interested colleagues. As is evident when the data are reviewed, there is a missing datum, due to difficulties encountered during field operations.

```
################################################################
# Housekeeping                         Use for All Analyses
################################################################
rm(list = ls())      # CAUTION: Remove all files in the working
                     # directory. If this action is not desired,
                     # use the rm() function one-by-one to remove
                     # the objects that are not needed.
setwd("F:/R_Biostatistics")
                     # Set to a new working directory.
                     # Note the single forward slash and double
                     # quotes.
                     # This new directory should be the directory
                     # where the data file is located, otherwise
                     # the data file will not be found.
getwd()              # Confirm the working directory.
search()             # Attached packages and objects.
################################################################
```

Create an object called PrePostWtUnstack.df. The object PrePostWtUnstack.df will be a dataframe, as indicated by the enumerated .df extension to the object name. This object will represent the output of applying the read.table() function against the comma-separated values file called Weight_PrePost_Supplement_Unstacked.csv. Note the arguments used with the read.table() function, showing that there is a header with descriptive variable names (header=TRUE) and that the separator between fields is a comma (sep=",").

```
PrePostWtUnstack.df <- read.table (file =
  "Weight_PrePost_Supplement_Unstacked.csv",
  header = TRUE,
  sep = ",")                # Import the .csv file

getwd()                     # Identify the working directory
ls()                        # List objects
attach(PrePostWtUnstack.df) # Attach the data, for later use
str(PrePostWtUnstack.df)    # Identify structure
nrow(PrePostWtUnstack.df)   # List the number of rows
```

```
ncol(PrePostWtUnstack.df)        # List the number of columns
dim(PrePostWtUnstack.df)         # Dimensions of the dataframe
names(PrePostWtUnstack.df)       # Identify names
colnames(PrePostWtUnstack.df)    # Show column names
rownames(PrePostWtUnstack.df)    # Show row names
head(PrePostWtUnstack.df)        # Show the head
tail(PrePostWtUnstack.df)        # Show the tail
PrePostWtUnstack.df              # Show the entire dataframe
summary(PrePostWtUnstack.df)     # Summary statistics
```

By completing this action, an object called PrePostWtUnstack.df has been created. This R-based object is a dataframe and it consists of the data originally included in the file Weight_PrePost_Supplement_Unstacked.csv, a comma-separated .csv file. To avoid possible conflicts, make sure that there are no prior R-based objects called PrePostWtUnstack.df. The prior use of rm(list=ls()) accommodates this concern, removing all prior objects in the current R session.

Note how it was only necessary to key the filename for the .csv file and not the full pathname since the R working directory is currently set to the directory and/or subdirectory where this .csv file is located (see the Housekeeping section at the beginning of this lesson).

Additionally, when showing the entire dataset, notice the listing for Subject 19 and how NA (the R notation for missing data) has been inserted for the KgPostSupplement datum. Review the help page for the read.table() function to see how blank fields, which applies for Subject 19, are viewed as missing values and that NA is inserted during execution of the read.table() function.

4.3 Organize the Data and Display the Code Book

For this lesson, the class() function, str() function, and duplicated() function will be used to be sure that data are organized as desired.

```
class(PrePostWtUnstack.df)
class(PrePostWtUnstack.df$Subject)
class(PrePostWtUnstack.df$KgPreSupplement)
class(PrePostWtUnstack.df$KgPostSupplement)

str(PrePostWtUnstack.df)                  # Structure

duplicated(PrePostWtUnstack.df)           # Duplicates
```

The class for each object seems to be correct and there are no duplicate rows of data in the dataframe. Saying this, a Code Book will help with future understanding of this dataset.

A Code Book is an essential aid for anyone involved in the day-to-day activities of the research and statistics process. The Code Book is typically brief and only

4.3 Organize the Data and Display the Code Book

serves as a useful reminder for what can be easily forgotten months (or even weeks) later, to make it easy to decipher what may otherwise be seen as arcane numeric codes. Coding schemes that are intuitive today can easily become obtuse tomorrow.

The Code Book below represents how data are desired before analyses begin. Recoding may be needed to put data into new formats.

```
########################################################
# Code Book                                             #
########################################################
#                                                      #
# Subject .................... Factor (e.g., nominal)  #
#                       A unique ID ranging from 1 to 150 #
#                                                      #
# KgPreSupplement     ....... Numeric (e.g., interval) #
#            Weight (Kg) of an unidentified biological #
#                specimen, ranging from 5.00 to 20.00  #
#                                                      #
# KgPostSupplement    ....... Numeric (e.g., interval) #
#            Weight (Kg) of an unidentified biological #
#                specimen, ranging from 5.00 to 20.00  #
########################################################
```

The str() function is then again applied against the dataframe to see the nature of each object variable as well as confirmation that the data are collectively viewed as a dataframe:

```
str(PrePostWtUnstack.df)
```

Recall that the Code Book shows data in their desired formats, which often requires some degree of recoding which has not yet occurred.

Once there is agreement that the data were brought into R in correct format, it is usually necessary to organize the data to some degree:

- The object variable Subject is currently viewed as an integer. Some R users may find it best to recode these identification-type numeric values into row names, but in this lesson Subject will instead be recoded into a factor object variable.
- Values for KgPreSupplement are viewed as numeric values.
- Values for KgPostSupplement are viewed as numeric values.

This transformation (again, typically called a recode action) is needed and the process, using R-based syntax, follows. There may be some unnecessary (perhaps redundant) actions with the following recode activities, but these are purposely done to provide assurance that each variable is in desired format:

```
PrePostWtUnstack.df$Subject           <- as.factor(
   PrePostWtUnstack.df$Subject)

PrePostWtUnstack.df$KgPreSupplement   <- as.numeric(
   PrePostWtUnstack.df$KgPreSupplement)
```

```
PrePostWtUnstack.df$KgPostSupplement     <- as.numeric(
  PrePostWtUnstack.df$KgPostSupplement)

getwd()                           # Identify the working directory
ls()                              # List objects
attach(PrePostWtUnstack.df)       # Attach the data, for later use
str(PrePostWtUnstack.df)          # Identify structure
nrow(PrePostWtUnstack.df)         # List the number of rows
ncol(PrePostWtUnstack.df)         # List the number of columns
dim(PrePostWtUnstack.df)          # Dimensions of the dataframe
names(PrePostWtUnstack.df)        # Identify names
colnames(PrePostWtUnstack.df)     # Show column names
rownames(PrePostWtUnstack.df)     # Show row names
head(PrePostWtUnstack.df)         # Show the head
tail(PrePostWtUnstack.df)         # Show the tail
PrePostWtUnstack.df               # Show the entire dataframe
summary(PrePostWtUnstack.df)      # Summary statistics
```

Note the formal nomenclature used in this recode action and the use of DataFrame$ObjectName notation when working with object variables that are part of a dataframe. Note also how the $ symbol is used to separate the name of the dataframe from the name of the object: DataFrame$ObjectName notation.

4.4 Conduct a Visual Data Check

Now that the data are all in proper format, it would be common to immediately apply the t-Test algorithm for matched pairs against the data and to complete this analysis. However, it is best to first prepare a few graphical displays of the data and to then reinforce comprehension of the data with descriptive statistics and measures of central tendency.

A few throwaway graphics will likely be sufficient at first, but fully-embellished graphical images may be needed later, for either publication or presentation. The full set of lessons, both prior lessons and future lessons, provide many examples of how graphics enhance understanding of the data.

```
par(ask=TRUE)
plot(PrePostWtUnstack.df$Subject,
  main="Plot of Subject")

par(ask=TRUE)
plot(density(PrePostWtUnstack.df$KgPreSupplement,
  na.rm=TRUE),    # Required for the density() function
  main="Density Plot of Pre-Supplement Weight (Kg)",
  lwd=6, col="red", font.axis=2, font.lab=2)
```

4.4 Conduct a Visual Data Check

```
par(ask=TRUE)
plot(density(PrePostWtUnstack.df$KgPostSupplement,
   na.rm=TRUE),    # Required for the density() function
   main="Density Plot of Post-Supplement Weight (Kg)",
   lwd=6, col="red", font.axis=2, font.lab=2)

par(ask=TRUE)
hist(PrePostWtUnstack.df$KgPreSupplement,
   main="Histogram of Pre-Supplement")

par(ask=TRUE)
hist(PrePostWtUnstack.df$KgPostSupplement,
   main="Histogram of Post-Supplement")
```

The throwaway graphics plot(), plot(density()), and hist(), along with other simple graphical functions, are all quite useful. As seen using the plot() function against PrePostWtUnstack.df$Subject, it seems that all subjects are included in the dataset. Further, the plot(density()) and hist() functions against the two numeric variables (PrePostWtUnstack.df$KgPreSupplement and PrePostWtUnstack.df$KgPostSupplement) generally follow normal distribution of values (e.g., approximate the bell-shaped curve). Saying this, a few more slightly embellished graphics will help better understand the nature of the data.

```
par(ask=TRUE)
boxplot(PrePostWtUnstack.df$KgPreSupplement,
   main="Boxplot of Weight (Kg) Pre-Supplement",
   xlab="Weight (Kg)",
   ylim=c(0,20), col="red", range=0)
box()
```

Note how PrePostWtUnstack.df$KgPreSupplement was used in this boxplot instead of KgPreSupplement. It is usually best to use DataFrame$ObjectName notation when naming object variables. Descriptive object names are desirable since they support internal documentation and subsequently promote good programming practice (gpp). Of course, the same naming approach applies for the object variable PrePostWtUnstack.df$KgPostSupplement.

```
par(ask=TRUE)
boxplot(PrePostWtUnstack.df$KgPostSupplement,
   main="Boxplot of Weight (Kg) Post-Supplement",
   xlab="Weight (Kg)",
   ylim=c(0,20), col="red", range=0)
box()

par(ask=TRUE)
boxplot(PrePostWtUnstack.df$KgPreSupplement,
```

```
  PrePostWtUnstack.df$KgPostSupplement,
  main="Boxplot of Weight (Kg) Pre-Supplement and
  Post Supplement",
  names=c("Pre-Supplement Weight (Kg)",
          "Post-Supplement Weight (Kg)"),
  ylim=c(0,20), col="red", range=0)
box()
```

Although it is useful to have separate boxplots, it is also useful to have the two distributions, as boxplots, displayed side-by-side. This action allows for a visual comparison of the distribution of both object variables.

Histograms also provide a general view of the data. The histograms below are slightly embellished.

```
par(ask=TRUE)
hist(PrePostWtUnstack.df$KgPreSupplement,
  main="Embellished Histogram of Weight (Kg)
  Pre-Supplement",
  xlab="Weight (Kg)", font.lab=2, font.axis=2,
  freq=TRUE, xlim=c(0,20), ylim=c(0,35), col="red")
box()

par(ask=TRUE)
hist(PrePostWtUnstack.df$KgPostSupplement,
  main="Embellished Histogram of Weight (Kg)
  Post-Supplement",
  xlab="Weight (Kg)", font.lab=2, font.axis=2,
  freq=TRUE, xlim=c(0,20), ylim=c(0,35), col="red")
box()

install.packages("UsingR")
library(UsingR)                 # Load the UsingR package.
help(package=UsingR)             # Show the information page.
sessionInfo()                    # Confirm all attached packages.

par(ask=TRUE)
UsingR::simple.hist.and.boxplot(
  PrePostWtUnstack.df$KgPreSupplement,
  main="Histogram, Rug, and Boxplot of Weight (Kg)
  Pre-Supplement")

par(ask=TRUE)
UsingR::simple.hist.and.boxplot(
  PrePostWtUnstack.df$KgPostSupplement,
  main="Histogram, Rug, and Boxplot of Weight (Kg)
  Post-Supplement")
```

4.4 Conduct a Visual Data Check

The UsingR::simple.hist.and.boxplot() function combines a histogram, rug, and a boxplot. Key help(rug) to learn more about this useful tool for another way of displaying data distributions.

```
install.packages("descr")
library(descr)              # Load the descr package.
help(package=descr)         # Show the information page.
sessionInfo()               # Confirm all attached packages.

savelwd     <- par(lwd=6)           # Heavy line
par(ask=TRUE)
descr:::histkdnc(PrePostWtUnstack.df$KgPreSupplement,
  breaks = 0, include.lowest = TRUE, right = TRUE,
  col = "wheat1", border="black",
  main =" Histogram of Weight (Kg) Pre-Supplement with
  Density Plot Overlay (Red) and Normal Curve (Blue)",
  xlab = "Weight (Kg)", font.lab=2, font.axis=2)
box()
par(savelwd) # Return to original value.

savelwd     <- par(lwd=6)           # Heavy line
par(ask=TRUE)
descr:::histkdnc(PrePostWtUnstack.df$KgPostSupplement,
  breaks = 0, include.lowest = TRUE, right = TRUE,
  col = "wheat1", border="black",
  main =" Histogram of Weight (Kg) Post-Supplement with
  Density Plot Overlay (Red) and Normal Curve (Blue)",
  xlab = "Weight (Kg)", font.lab=2, font.axis=2)
box()
par(savelwd) # Return to original value.
```

The descr:::histkdnc() function generates an attractive and concise display, enhancing visualization of data distribution.

```
par(ask=TRUE)
Pre.histogram <- hist(PrePostWtUnstack.df$KgPreSupplement)

par(ask=TRUE)
plot(Pre.histogram,
  main="Histogram of Pre-Supplement Weight (Kg)
  with Added Frequency Values",
  xlab="Weight (Kg)", font.lab=2, font.axis=2,
  freq=TRUE, xlim=c(0,20), ylim=c(0,35), col="red")
text(Pre.histogram$mids,
  Pre.histogram$counts+2,
  label=c(Pre.histogram$counts), font=2)
box()

par(ask=TRUE)
Post.histogram <- hist(PrePostWtUnstack.df$KgPostSupplement)
```

```
par(ask=TRUE)
plot(Post.histogram,
  main="Histogram of Post-Supplement Weight (Kg)
  with Added Frequency Values",
  xlab="Weight (Kg)", font.lab=2, font.axis=2,
  freq=TRUE, xlim=c(0,20), ylim=c(0,35), col="red")
text(Post.histogram$mids,
  Post.histogram$counts+2,
  label=c(Post.histogram$counts), font=2)
box()
```

In the above example, a histogram is first prepared for the selected object variable. Then, text is added to the histogram, showing the number of values (e.g., count, frequency values, etc.) for each bar.

```
install.packages("Hmisc")
library(Hmisc)              # Load the Hmisc package.
help(package=Hmisc)         # Show the information page.
sessionInfo()               # Confirm all attached packages.

savefont      <- par(font=2)       # Bold text
savefont.lab  <- par(font.lab=2)   # Bold labels
savefont.axis <- par(font.axis=2)  # Bold axis
par(Ask=TRUE)
Side_by_Side_Histogram <- Hmisc::histbackback(
  PrePostWtUnstack.df$KgPreSupplement,
  PrePostWtUnstack.df$KgPostSupplement,
  xlab=c("Pre-Supplement Weight (Kg)",
         "Post-Supplement Weight (Kg)"),
  ylab="Weight (Kg)", probability=TRUE, axes=TRUE,
  xlim=c(-0.25, 0.25),
  main="Pre-Supplement Weight (Kg) v Post-Supplement
  Weight (Kg)")
barplot(-Side_by_Side_Histogram$left,   # See the - sign
  col="red" , horiz=TRUE, space=0, add=TRUE, axes=FALSE)
barplot(Side_by_Side_Histogram$right,
  col="blue", horiz=TRUE, space=0, add=TRUE, axes=FALSE)
legend("topleft",
  bty="n", fill="red", "Pre-Supplement     ")
legend("bottomright",
  bty="n", fill="blue", "Post-Supplement    ")
par(savefont)        # Return to default setting
par(savefont.lab)    # Return to default setting
par(savefont.axis)   # Return to default setting
box()
```

A set of side-by-side histograms is a powerful tool for visualizing the distribution of two numeric object vectors. The Hmisc::histbackback() function is one of the best tools for this purpose.

4.4 Conduct a Visual Data Check

Overlapping density plots are another useful tool for displaying data distribution of two separate object variables. Of course, the selected inferential analysis (Student's t-Test for Matched Pairs in this lesson) is used to determine if there is a statistically significant difference between the two distributions.

```
par(ask=TRUE)
plot(density(PrePostWtUnstack.df$KgPreSupplement,
  na.rm=TRUE),
  main="Density Plot of Pre-Supplement Weight (Kg) and
  Post Supplement Weight (Kg)",
  xlab="Weight (Kg)", font.lab=2, font.axis=2, xlim=c(0,20),
  ylim=c(0,0.22),          # Adjust to meet the specific data
  lwd=8, lty=1, col="red")
lines(density(PrePostWtUnstack.df$KgPostSupplement,
  na.rm=TRUE),
  font.lab=2, font.axis=2,
  xlim=c(0,20),            # Adjust to meet the specific data
  lwd=8, lty=6, col="blue")
savefont <- par(font=2)        # Bold text
legend("topleft",
  bty="n", fill="red", "Pre-Supplement Weight (Kg)       ")
legend("topright",
  bty="n", fill="blue", "Post-Supplement Weight (Kg)      ")
par(savefont)                  # Return to default setting
box()
```

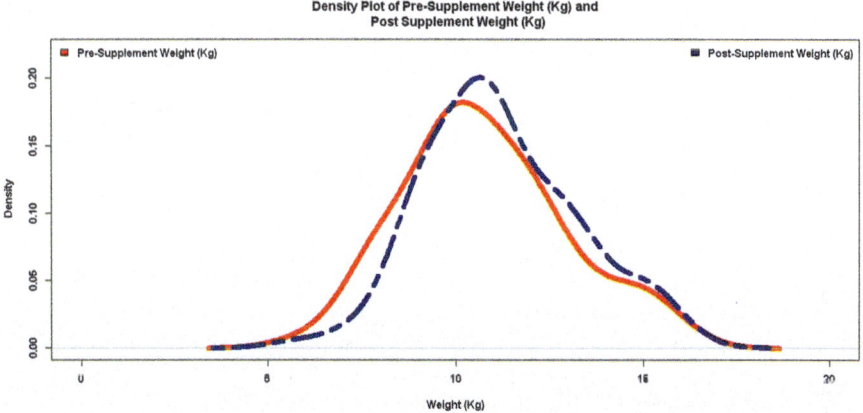

In this figure, the distribution of values for the two object variables KgPreSupplement and KgPostSupplement is not an exact facsimile of values in a normal curve, but the distribution is adequate and these graphical images help in judging the appropriate use of the parametric Student's t-Test for Matched Pairs, where normal distribution is desired, but rarely achieved at the level of perfection.

4.5 Descriptive Analysis of the Data

This dataset introduces the need for attention to missing data. Given the different ways missing data can impact analyses, it is often helpful to first check for missing data by using the is.na() function and the complete.cases() function against the entire dataset. Both functions return a TRUE or FALSE response, depending on the function and the outcome of whether data are missing or data are not missing.

```
is.na(PrePostWtUnstack.df)              # Check for missing data
complete.cases(PrePostWtUnstack.df)     # Check for complete cases
```

For the dataset PrePostWtUnstack.df note how there is a missing KgPostSupplement datum for Subject 19 and of course, this event shows when using the is.na() function and the complete.cases() function.

For this simple dataset, the summary() function may be all that is necessary to gain a sense of the data. Note how the summary() function is applied against the entire dataset, thus yielding information about all object variables, including the object variable Subject.

```
summary(PrePostWtUnstack.df)
```

When viewing the summary() function output, give attention to the listing of NA for KgPostSupplement. Again, the summary() function is very useful and it should always be a first selection when preparing descriptive analyses.

Although the summary() function is often sufficient, descriptive statistics for individual object variables may be desired. To achieve this aim, review the prior lesson *Descriptive Statistics and Measures of Central Tendency* for a comprehensive review of the functions used for descriptive statistics, especially: length(), asbio::Mode(), median(), mean(), sd(), table(), and finally summary(). As needed (but not always, depending on specific functions), the na.rm=TRUE argument or some other similar convention will be used to accommodate missing data.

```
length(PrePostWtUnstack.df$KgPreSupplement)      # N
length(PrePostWtUnstack.df$KgPostSupplement)     # N

install.packages("asbio")
library(asbio)              # Load the asbio package.
help(package=asbio)         # Show the information page.
sessionInfo()               # Confirm all attached packages.

asbio::Mode(PrePostWtUnstack.df$KgPreSupplement)     # Mode
asbio::Mode(PrePostWtUnstack.df$KgPostSupplement)    # Mode
```

The asbio::Mode() function does not easily accommodate missing values, if at all. Given this challenge, a simple use of R syntax should be all that is needed to hand-calculate mode, the most frequently occurring value in an array of values.

```
mode.of.KgPreSupplement <-
  names(sort(-table(PrePostWtUnstack.df$KgPreSupplement)))[1]
mode.of.KgPreSupplement      # Hand calculate mode with NAs
```

4.5 Descriptive Analysis of the Data

```
mode.of.KgPostSupplement <-
  names(sort(-table(PrePostWtUnstack.df$KgPostSupplement)))[1]
mode.of.KgPostSupplement    # Hand calculate mode with NAs

median(PrePostWtUnstack.df$KgPreSupplement,
  na.rm=TRUE)                                              # Median
median(PrePostWtUnstack.df$KgPostSupplement,
  na.rm=TRUE)                                              # Median

mean(PrePostWtUnstack.df$KgPreSupplement, na.rm=TRUE)  # Mean
sd(PrePostWtUnstack.df$KgPreSupplement,na.rm=TRUE )    # SD
  # Measures of Central Tendency

mean(PrePostWtUnstack.df$KgPostSupplement,na.rm=TRUE )# Mean
sd(PrePostWtUnstack.df$KgPostSupplement, na.rm=TRUE)  # SD
  # Measures of Central Tendency

summary(PrePostWtUnstack.df)
```

The epicalc::summ() function is also useful in that it can provide descriptive statistics and a representative figure of individual object variables.

```
install.packages("epicalc")
library(epicalc)          # Load the epicalc package.
help(package=epicalc)     # Show the information page.
sessionInfo()             # Confirm all attached packages.

par(ask=TRUE) # Use the epicalc package.
epicalc::summ(PrePostWtUnstack.df$KgPreSupplement,
  by=NULL, graph=TRUE, box=TRUE,   # Generate a boxplot
  pch=18, ylab="auto",
  main="Sorted Dotplot and Boxplot of
  Pre-Supplement Weight (Kg)",
  cex.X.axis=1.15, cex.Y.axis=1.15, font.lab=2,
  dot.col="auto")
  # Note the descriptive statistics that go
  # along with the  epicalc::summ() function.

par(ask=TRUE) # Use the epicalc package.
epicalc::summ(PrePostWtUnstack.df$KgPostSupplement,
  by=NULL, graph=TRUE, box=TRUE,   # Generate a boxplot
  pch=18, ylab="auto",
  main="Sorted Dotplot and Boxplot of
  Pre-Supplement Weight (Kg)",
  cex.X.axis=1.15, cex.Y.axis=1.15, font.lab=2,
  dot.col="auto")
  # Note the descriptive statistics that go
  # along with the  epicalc::summ() function.
```

Although the epicalc::summ() function may be sufficient for production of descriptive statistics, there are many other functions that serve the same purpose.

A few will be demonstrated, including the prettyR::describe() function, the psych::describe() function, and the lessR::SummaryStats() function. As time permits, explore the many other R functions that serve a similar purpose.

```
install.packages("prettyR")
library(prettyR)              # Load the prettyR package.
help(package=prettyR)         # Show the information page.
sessionInfo()                 # Confirm all attached packages.

prettyR::describe(PrePostWtUnstack.df$KgPreSupplement)
prettyR::describe(PrePostWtUnstack.df$KgPostSupplement)

install.packages("psych")
library(psych)                # Load the psych package.
help(package=psych)           # Show the information page.
sessionInfo()                 # Confirm all attached packages.

psych::describe(PrePostWtUnstack.df$KgPreSupplement)
psych::describe(PrePostWtUnstack.df$KgPostSupplement)

install.packages("lessR")
library(lessR)                # Load the lessR package.
help(package=lessR)           # Show the information page.
sessionInfo()                 # Confirm all attached packages.

lessR::SummaryStats(KgPreSupplement, data=PrePostWtUnstack.df)
lessR::SummaryStats(KgPostSupplement, data=PrePostWtUnstack.df)
```

The tables::tabular() function can be used to provide even more detail, in a fairly attractive table format that can be easily copied or used in some other fashion in a summary report.

```
install.packages("tables")
library(tables)               # Load the tables package.
help(package=tables)          # Show the information page.
sessionInfo()                 # Confirm all attached packages.

tables::tabular(KgPreSupplement*(length + min + max + mean) ~
  1, data=PrePostWtUnstack.df)
tables::tabular(KgPostSupplement*(length + min + max + mean) ~
  1, data=PrePostWtUnstack.df )
```

The tables::tabular() function cannot immediately accommodate missing values, such as the case for Subject 19 – KgPostSupplement. A simple enumeration of a new set of functions for length, min, max, and mean will take care of this concern.

```
LENGTH <- function(x) base::length(x) # Redundant for length
MIN    <- function(x) base::min(x, na.rm=TRUE)
MAX    <- function(x) base::max(x, na.rm=TRUE)
MEAN   <- function(x) base::mean(x, na.rm=TRUE)

tables::tabular(KgPreSupplement*(length + min + max + mean) ~
  1, data=PrePostWtUnstack.df)
```

4.6 Conduct the Statistical Analysis

```
tables::tabular(KgPostSupplement*(LENGTH + MIN + MAX + MEAN) ~
  1, data=PrePostWtUnstack.df )    # Note the use of CAPS here
```

Additional functions could be demonstrated, but the above functions should provide a broad representation of how descriptive statistics and measures of central tendency are determined when using R. Our immediate concern for the two object variables in question is again easily viewed by using the summary() function:

```
> summary(PrePostWtUnstack.df)
    Subject      KgPreSupplement   KgPostSupplement
1       : 1     Min.    : 5.57    Min.    : 5.57
2       : 1     1st Qu. : 9.28    1st Qu. : 9.64
3       : 1     Median  :10.60    Median  :10.89
4       : 1     Mean    :10.79    Mean    :11.19
5       : 1     3rd Qu. :12.24    3rd Qu. :12.56
6       : 1     Max.    :16.47    Max.    :16.30
(Other) :144                      NA's    :1
```

With sufficient experience, preferences and individual choice will help determines which functions to use. For now it is only necessary to determine if there is a statistically significant difference in KgPreSupplement as compared to KgPostSupplement.

4.6 Conduct the Statistical Analysis

The dataset for this part of the lesson was prepared in unstacked format. A brief demonstration of the same data, but presented as a dataset in stacked format, follows later in this lesson. Review the help pages for the stack() function and the unstack() function to learn more about this issue.

Although a great deal of information has been given to visual presentations of data distribution and descriptive statistics, our primary emphasis for this lesson revolves around Student's t-Test for Matched Pairs.

Use the t.test() function and note in the two examples, below, differences in the way variance is viewed: default selection, declared equal variance, declared unequal variance.

....

```
t.test(PrePostWtUnstack.df$KgPreSupplement,   # Measured variable
   PrePostWtUnstack.df$KgPostSupplement,      # Measured variable
   paired=TRUE,                               # Matched pairs
   na.rm=TRUE)                                # Missing data

Outcome:   t = -7.5467, df = 148, p-value = 0.000000000004206
```
....

```
t.test(PrePostWtUnstack.df$KgPreSupplement,   # Measured variable
  PrePostWtUnstack.df$KgPostSupplement,       # Measured variable
  alternative="two.sided",                    # Two-sided t-Test
  paired=TRUE,                                # Matched pairs
  na.rm=TRUE,                                 # Missing data
  var.equal=TRUE)                             # Equal variance

Outcome:    t = -7.5467, df = 148, p-value = 0.000000000004206
```

```
t.test(PrePostWtUnstack.df$KgPreSupplement,   # Measured variable
  PrePostWtUnstack.df$KgPostSupplement,       # Measured variable
  alternative="two.sided",                    # Two-sided t-Test
  paired=TRUE,                                # Matched pairs
  na.rm=TRUE,                                 # Missing data
  var.equal=FALSE)                            # Equal variance
                                              # is not TRUE
Outcome:    t = -7.5467, df = 148, p-value = 0.000000000004206
```

Notice how a comma and not a tilde was used to separate the unstacked objects PrePostWtUnstack.df$KgPreSupplement and PrePostWtUnstack.df$KgPostSupplement. There is no grouping (e.g., binary) variable with unstacked data as there is with stacked data since the two sets of unstacked data represent their own groups. By design, look at the error message below, which was generated by R when a ~ instead of a , was used with unstacked data.

```
t.test(PrePostWtUnstack.df$KgPreSupplement ~  # Mesured variable
  PrePostWtUnstack.df$KgPostSupplement,       # Measred variable
  paired=TRUE,                                # Matched pairs
  na.rm=TRUE)                                 # Missing data

Outcome:    Error grouping factor must have exactly 2 levels
```

Key help(t.test) at the R prompt to learn more about the many arguments associated with the t.test() function. Of special importance for many circumstances is the issue of the declared confidence interval.

As demonstrated below, a 95 % confidence interval is the default confidence for the t.test() function. Note how it was not necessary to declare this argument in the following syntax:

```
# No declared confidence level - use default
t.test(PrePostWtUnstack.df$KgPreSupplement,
  PrePostWtUnstack.df$KgPostSupplement,
  alternative="two.sided",
  paired=TRUE, na.rm=TRUE, var.equal=TRUE)
```

Outcome: 95 percent confidence interval:
           -0.4829758 -0.2825275

There may be times, however, when a different confidence interval is needed. To achieve that aim, merely use the conf= argument, as demonstrated below:

```
# Adjust to 90 percent confidence interval.
t.test(PrePostWtUnstack.df$KgPreSupplement,
  PrePostWtUnstack.df$KgPostSupplement,
  alternative="two.sided",
  paired=TRUE, na.rm=TRUE,
  conf=0.9,                # Not default selection
  var.equal=TRUE)
```

Outcome: 90 percent confidence interval:
           -0.4667002 -0.2988032

As useful as the t.test() function is, it is often a good idea to analyze the same dataset with a complementary function, as a redundant data check. For comparison of statistics generated by the t.test() function, consider the lessR::ttest() function. Although the interface for how outcomes are presented with the lessR::ttest() function will be different than how outcomes show with the t.test() function, the statistics should be equivalent, with the possible observation of rounding.

```
# Use of the lessR::ttest() function as a redundant
# check on the t.test() function
lessR::ttest(KgPreSupplement, KgPostSupplement,
  data=PrePostWtUnstack.df, paired=TRUE)
```

Outcome: t-value = -7.547, df = 148, p-value = 0.000

As with all other R functions, first learn the basics and work with default settings. Then use the help pages and other learning resource materials. RSeek (http://rseek.org) is especially helpful to initiate search strategies.

4.7 Summary

In this lesson, the graphics and statistics provided a great deal of information. Of immediate importance, however, focus on outcomes of the Student's t-Test for Matched Pairs, used to address the Null Hypothesis.

Assuming that the two variances are equal, the following statistics of importance have been gained from two sets of analyses (descriptive statistics and Student's t-Test) for this problem:

```
                    n    mean   sd
KgPreSupplement    150  10.79  2.22
KgPostSupplement   149  11.19  2.07
```

```
t = -7.5467, df = 148, p-value = 0.000000000004206
```

Attention to the p-value is perhaps the easiest way to view differences for this problem:

- The calculated p-value is 0.000000000004206.
- The declared p-value, from the previously stated Null Hypothesis, is 0.05.
- The calculated p-value is less than the declared p-value.

Therefore, the Null Hypothesis is rejected (e.g., not accepted) and it can be claimed that there is a difference (p <= 0.05) between KgPreSupplement and KgPostSupplement. The weight (Kg) for subjects was greater after the supplement was provided than the weight (Kg) of subjects before the supplement was provided and this difference is not due to chance but instead represents a true difference. Whether it was intended or not, subjects gained weight after application of the treatment (e.g., introduction of a mineral supplement into a maintenance feeding program).

There is no knowledge if gender, breed, or any other factor-type variable had any possible influence of outcomes. Data were not provided for these factors. Instead, as this pre-test and post-test repeated measures lesson was structured, it is only known that matched subjects, overall at p <= 0.05, had a significant difference in weight after administration of the feeding supplement compared to weights before administration of the supplement.

As an additional comment, know that in some texts, the phrase *The null hypothesis was rejected.* may instead be stated as *The null hypothesis was not accepted.* but discussion on differences in the fine points of these two phrases is beyond the purpose of this introductory lesson.

4.8 Addendum 1: Stacked Data and Student's t-Test for Matched Pairs

For this lesson, the data have been organized into two separate datasets, to demonstrate how R accommodates both unstacked data (the main body of this lesson) and stacked data (this addendum). Both datasets, however, have been prepared in .csv (e.g., comma-separated values) file format. The data are separated by commas. The data are not separated by tabs and the data are not separated by spaces.

In this addendum, the same data that were previously presented in unstacked format are now brought into R in stacked format. To avoid excessive redundancy only a few analyses will be presented, with emphasis on how Student's t-Test for Matched Pairs is organized with stacked data.

Create an object called PrePostWtStack.df. The object PrePostWtStack.df will be a dataframe, as indicated by the enumerated .df extension to the object name. This object will represent the output of applying the read.table() function against the comma-separated values file called Weight_PrePost_Supplement_Stacked.csv. Note the arguments used with the read.table() function, showing that there is a

4.8 Addendum 1: Stacked Data and Student's t-Test for Matched Pairs

header with descriptive variable names (header = TRUE) and that the separator between fields is a comma (sep = ",").

```
PrePostWtStack.df <- read.table (file =
  "Weight_PrePost_Supplement_Stacked.csv",
  header = TRUE,
  sep = ",")                    # Import the .csv file

getwd()                         # Identify the working directory
ls()                            # List objects
attach(PrePostWtStack.df)       # Attach the data, for later use
str(PrePostWtStack.df)          # Identify structure
nrow(PrePostWtStack.df)         # List the number of rows
ncol(PrePostWtStack.df)         # List the number of columns
dim(PrePostWtStack.df)          # Dimensions of the dataframe
names(PrePostWtStack.df)        # Identify names
colnames(PrePostWtStack.df)     # Show column names
rownames(PrePostWtStack.df)     # Show row names
head(PrePostWtStack.df)         # Show the head
tail(PrePostWtStack.df)         # Show the tail
PrePostWtStack.df               # Show the entire dataframe
summary(PrePostWtStack.df)      # Summary statistics
```

By completing this action, an object called PrePostWtStack.df has been created. This R-based object is a dataframe and it consists of the data originally included in the file Weight_PrePost_Supplement_Stacked.csv, a comma-separated .csv file. To avoid possible conflicts, make sure that there are no prior R-based objects called PrePostWtStack.df.

For this lesson, the class() function, str() function, and duplicated() function will be used to be sure that data are organized as desired.

```
class(PrePostWtStack.df)
class(PrePostWtStack.df$Pair)
class(PrePostWtStack.df$TimePeriod)
class(PrePostWtStack.df$Kg)

str(PrePostWtStack.df)              # Structure

duplicated(PrePostWtStack.df)       # Duplicates
```

The class for each object seems to be correct and there are no duplicate rows of data in the dataframe. Saying this, a Code Book will help with future understanding of this dataset.

```
########################################################
# Code Book                                             #
########################################################
#                                                      #
# Pair      .................. Factor (e.g., nominal) #
```

```
#                         A unique ID ranging from 1 to 150 #
#                                                           #
# TimePeriod    ..............  Factor (e.g., nominal)      #
#                                              PreSupplement #
#                                             PostSupplement #
#                                                           #
# Kg            ....................  Numeric (e.g., interval) #
#              Weight (Kg) of an unidentified biological    #
#              specimen, ranging from 5.00 to 20.00         #
#############################################################
```

The str() function is then again applied against the dataframe to see the nature of each object variable as well as confirmation that the data are collectively viewed as a dataframe:

```
str(PrePostWtStack.df)
```

Once there is agreement that the data were brought into R in correct format, it is usually necessary to organize the data to some degree:

- The object variable Pair is currently viewed as an integer. Some R users may find it best to recode these identification-type numeric values into row names, but in this lesson Pair will instead be recoded into a factor object variable.
- Values for TimePeriod are in alpha-format and are viewed as factor values.
- Values for Kg are viewed as numeric values.

This transformation (again, typically called a recode action) is needed and the process, using R-based syntax, follows. There may be some unnecessary (perhaps redundant) actions with the following recode activities, but these are purposely done to provide assurance that each variable is in desired final format:

```
PrePostWtStack.df$Pair           <- as.factor(
   PrePostWtStack.df$Pair)

PrePostWtStack.df$TimePeriod     <- as.factor(
   PrePostWtStack.df$TimePeriod)

PrePostWtStack.df$Kg             <- as.numeric(
   PrePostWtStack.df$Kg)

getwd()                          # Identify the working directory
ls()                             # List objects
attach(PrePostWtStack.df)        # Attach the data, for later use
str(PrePostWtStack.df)           # Identify structure
nrow(PrePostWtStack.df)          # List the number of rows
ncol(PrePostWtStack.df)          # List the number of columns
dim(PrePostWtStack.df)           # Dimensions of the dataframe
names(PrePostWtStack.df)         # Identify names
colnames(PrePostWtStack.df)      # Show column names
rownames(PrePostWtStack.df)      # Show row names
```

4.8 Addendum 1: Stacked Data and Student's t-Test for Matched Pairs

```
head(PrePostWtStack.df)        # Show the head
tail(PrePostWtStack.df)        # Show the tail
PrePostWtStack.df              # Show the entire dataframe
summary(PrePostWtStack.df)     # Summary statistics
```

Data in this active R session are in two formats, originally as unstacked data (PrePostWtUnstack.df) and also as stacked data (PrePostWtStack.df). It is best to check the data to be sure that everything is correct. The summary() function may be the easiest way to see if data are equivalent.

```
# Check data in the unstacked dataset
summary(PrePostWtUnstack.df$KgPostSupplement)
summary(PrePostWtUnstack.df$KgPreSupplement)


# Check data in the stacked dataset, using the tapply()
# function to force breakout analyses
tapply(Kg, TimePeriod, summary, na.rm=TRUE,
   data=PrePostWtStack.df)
   # Breakouts of Kg by TimePeriod (e.g.,
   # PreSupplement or PostSupplement)
```

The two datasets are equivalent, as expected. Given this equivalence, apply the t.test() function against both datasets to confirm that inferential outcomes are also equivalent.

```
# Student's t-Test for matched pairs, against unstacked data
# Default selections.
t.test(PrePostWtUnstack.df$KgPreSupplement,  # Measured variable
   PrePostWtUnstack.df$KgPostSupplement,     # Measured variable
   paired=TRUE,                              # Matched pairs
   na.rm=TRUE)                               # Missing data

Outcome:   t = -7.5467, df = 148, p-value = 0.000000000004206

# Student's t-Test for matched pairs, against stacked data
# Default selections.
t.test(PrePostWtStack.df$Kg ~                # Measured variable
   PrePostWtStack.df$TimePeriod,             # Grouping variable
   paired=TRUE,                              # Matched pairs
   na.rm=TRUE)                               # Missing data

ERROR MESSAGE:  not all arguments have the same length
```

The error is of course generated because there is a missing PostSupplement datum for Pair 19. To accommodate this problem of unequal length, a simple *kludge* is used, where Pair 19 is deleted from the PrePostWtStack.df dataset. Because of this action there will now be an equal number of comparisons (e.g., pairs) between PreSupplement and PostSupplement.

Show the original dataset (PrePostWtStack.df), with all rows.

```
PrePostWtStack.df
nrow(PrePostWtStack.df)         # List the number of rows.
```

Determine which rows in PrePostWtStack.df$Pair represent Pair 19, the pair responsible for the missing data.

```
which(PrePostWtStack.df$Pair == "19")
```

Create a new dataset (NewPrePostWtStack.df), eliminating row 19 and row 169 since these are the two rows that have data for Pair 19.

```
NewPrePostWtStack.df <- PrePostWtStack.df[c(-19,-169),]
```

Show the new dataset (NewPrePostWtStack.df), which should not have data for Pair 19.

```
NewPrePostWtStack.df
nrow(NewPrePostWtStack.df)      # List the number of rows
```

Use the newly created dataset NewPrePostWtStack.df to conduct the Student's t-Test for matched pairs using stacked data, but now with no concern about unequal lengths (e.g., missing data for one of the pairs).

```
# Student's t-Test for matched pairs, against stacked data
t.test(NewPrePostWtStack.df$Kg ~      # Measured variable
   NewPrePostWtStack.df$TimePeriod,   # Grouping variable
   paired=TRUE,                       # Matched pairs
   na.rm=TRUE)                        # Missing data

Outcome:   t = -7.5467, df = 148, p-value = 0.000000000004206
```

There may be more elegant ways to resolve the problem of how missing data impact Student's t-Test for Matched Pairs, but this action works and it also showed how data can be managed to meet needs, which was a secondary purpose for showing this approach.

4.9 Addendum 2: The Impact of N on Student's t-Test

The results of the Student's t-Test for Matched Pairs provided evidence that there was a statistically significant difference ($p <= 0.5$) in Pre-Supplement Weight (Kg) and Post-Supplement Weight (Kg) for the entire dataset of 150 subjects, where individual biological specimens were matched against their weight change (if any) after a supplement of some type was introduced into the feeding program:

```
                 n    mean    sd
KgPreSupplement  150  10.79   2.22
KgPostSupplement 149  11.19   2.07

Outcome:   t = -7.5467, df = 148, p-value = 0.000000000004206
```

4.9 Addendum 2: The Impact of N on Student's t-Test

Think of the cost (actual dollars and time-on-task) for managing a research study with 150 biological subjects. Going beyond the initial purchase cost, these specimens need housing, feed, water, ventilation, exercise, seven-day-a-week daily care and management, etc. The cost of care for biological specimens is by no means insignificant. Is it possible to obtain the same general outcomes with a smaller, and therefore less expensive, sample of 30 subjects? To explore this question, obtain the base statistics from the original sample, then use the sample() function against the dataframe to obtain a sample of only 30 subjects.

```
head(PrePostWtUnstack.df)                              # N = 150
tail(PrePostWtUnstack.df)                              # N = 150
psych::describe(PrePostWtUnstack.df)                   # N = 150

# Apply the sample() function
Sample30.PrePostWtUnstack.df <- PrePostWtUnstack.df[
  sample(1:dim(PrePostWtUnstack.df)[1],
  size=30, replace=FALSE),]

head(Sample30.PrePostWtUnstack.df)                     # N = 30
tail(Sample30.PrePostWtUnstack.df)                     # N = 30
psych::describe(Sample30.PrePostWtUnstack.df)          # N = 30

# ---> N = 150 < ---
# Assumption that the two variances are equal.
t.test(PrePostWtUnstack.df$KgPreSupplement,
  PrePostWtUnstack.df$KgPostSupplement,
  alternative="two.sided",
  paired=TRUE, na.rm=TRUE, var.equal=TRUE)

# ---> N =  30 < ---
# Assumption that the two variances are equal.
t.test(Sample30.PrePostWtUnstack.df$KgPreSupplement,
  Sample30.PrePostWtUnstack.df$KgPostSupplement,
  alternative="two.sided",
  paired=TRUE, na.rm=TRUE, var.equal=TRUE)
```

Observe the p-value for when N = 150 and then observe the p-value for when N = 30. Use the sample() function multiple times, to see if results are generally consistent. Results will depend on the 30 subjects included in the selected sample so of course the p-value will change for each sample of 30 subjects.

Typically, as seen in this sample, the general outcome is upheld so that for both N = 150 and N = 30, there is a statistically significant difference (p <= 0.05) in Pre-Supplement Weight (Kg) and the matched Post-Supplement Weight (Kg). And of course, a sample of N = 30 would be far less expensive and subsequently easier to manage than a sample of N = 150. Research and statistical analyses are as much about budgets and the prudent management of resources as they are about generated p-values and graphical output.

Although there are those who think it is impressive to focus on gaining a large sample, it is perhaps more important to focus on the issue of sample representation. Is the above iteration of 30 subjects representative of the collection of all 150 subjects? That issue is beyond the purpose of this lesson, but sample representation should always be considered when conducting statistical analyses, regardless of the computing platform or selected software.

4.10 Prepare to Exit, Save, and Later Retrieve This R Session

It is common to prepare R syntax in a separate file, using a simple ASCII text editor. If time permits, experiment with Crimson Editor, Tinn-R, or vim, but there are many other possible selections.

Use the following set of actions to exit from the current R session.

```
getwd()             # Identify the current working directory.
ls()                # List all objects in the working
                    # directory.
ls.str()            # List all objects, with finite detail.
list.files()        # List files at the PC directory.

save.image("R_Lesson_t-Test_MatchedPairs.rdata")

getwd()             # Identify the current working directory.
ls()                # List all objects in the working
                    # directory.
ls.str()            # List all objects, with finite detail.
list.files()        # List files at the PC directory.

alarm()             # Alarm, notice of upcoming action.
q()                 # Quit this session.
                    # Prepare for Save workspace image? query.
```

Use the R Graphical User Interface (GUI) to load the saved rdata file: **File** and then **Load Workspace**. Otherwise, use the load() function, keying the full pathname, to load the .rdata file and retrieve the session.

Recall, however, that it may be just as useful to simply use a R script file (typically saved as a .txt ASCII-type file) and recreate the analyses and graphics, provided the data files remain available.

Chapter 5
Oneway Analysis of Variance (ANOVA)

Abstract The purpose of this lesson is to provide guidance on how R can be used to compare differences to a singular measured variable between three or more groups using Oneway Analysis of Variance (ANOVA). Along with instruction on the use of R and R syntax associated with Oneway ANOVA, this lesson will also reinforce the use of descriptive statistics and graphical figures to complement outcomes from the parametric Oneway ANOVA.

Keywords ANOVA • Analysis of Variance • Duncan's multiple range test • Least Significant Difference (LSD) • Mean comparison technique • Oneway ANOVA • Student-Newman-Keuls (SNK) • Tukey's Honestly Significant Difference (Tukey HSD) • Scheffé

5.1 Background on This Lesson

5.1.1 Description of the Data

This lesson on Oneway Analysis of Variance (ANOVA) is used to determine if there are differences in measurements of a biological specimen by a factor variable with five breakout groups. The exact nature of the dataset is not provided in this lesson. The focus of this lesson is on two specific variables, a factor-type variable with five breakout groups (F2b) and a numeric variable (M1).

For this lesson, assume that data are from biological specimens of the same type and that there are data from more than 4,000 subjects:

- Perhaps the data are from dairy cattle and the five breakout groups are dairy cattle breeds (Ayrshire, Brown Swiss, Guernsey, Holstein, and Jersey) while the measurement in question for this sample is some measure of milk production at end of lactation for first lactation.

- Perhaps the biological specimen represents hay as a forage crop and the five breakout are specific types of hay (alfalfa, clover, fescue, orchard grass, and timothy) while the measurement in question for this sample is some measure of protein for first cut hay.
- Perhaps the data are from humans and the five breakout groups are for race-ethnicity classifications in the United States (American Indian or Alaskan Native, Asian or Pacific Islander, Black non-Hispanic, Hispanic, and White non-Hispanic) while the measurement in question is some measure of blood pressure after 15 min of exercise on a treadmill.

A common statistical technique to determine if differences exist between three or more groups (however the concept of group is defined) is Oneway Analysis of Variance (ANOVA). ANOVA methodology involves the determination of differences for either:

- One group with multiple (typically, three or more) variations
- One variable, compared to multiple (typically, three or more) groups

For this lesson, merely assume that subjects are all of the same genus and species and that the measurement in question (M1) is both reliable and valid. The measurement in question will then be viewed by the five breakout groups (F2b) of the factor in question. Oneway ANOVA (Analysis of Variance) was correctly judged to be the appropriate test for this analysis of summative differences in measurement of M1 by the five breakout groups for F2b.

When using Oneway ANOVA for three or more groups, an immediate concern is how to interpret findings if the Null Hypothesis is not accepted. When only two groups are compared (such as the case when Student's t-Test is used) and if the Null Hypothesis is not accepted, then it is known that the difference between Group A and Group B is a true difference (at the declared level of significance, or p levels such as `p <= 0.05` or `p <= 0.01`).

What happens, however, if the Null Hypothesis for Oneway ANOVA with three groups is rejected? There is a known difference in group means, but where?

Comparisons, for Group A, Group B, and Group C could possibly include the following:

- Is the difference between Group A and Group B (or Group B and A, depending on ordering) the only reason for failure to accept the Null Hypothesis?
- Is the difference between Group A and Group C (or Group C and A, depending on ordering) the only reason for failure to accept the Null Hypothesis?
- Is the difference between Group B and Group C (or Group C and B, depending on ordering) the only reason for failure to accept the Null Hypothesis?
- Is there a difference in all three groups, where Group A is separate from all others, Group B is separate from all others, and Group C is separate from all others?

Oneway ANOVA designs can be quite complex but discussion on research design is beyond the scope of this lesson. Instead, it is only necessary to mention that Oneway ANOVA calls for careful attention to breakout statistics for all groups and of course, complexity is only magnified as the number of groups increases beyond three.

Given how there are multiple comparisons to consider with Oneway ANOVA, give attention to the following mean comparison tests, which are commonly used for the purpose of comparing differences between means in a Oneway ANOVA:

Oneway ANOVA mean comparison techniques	
Duncan	Duncan's multiple range test
LSD	Least-Significant Difference
Scheffé	Scheffé's test
SNK	Student-Newman-Keuls
Tukey	Tukey's Honestly Significant Difference (HSD)

Statistical tests that can account for increased complexity are needed if meaningful decisions involving statistics in the large are to be effected. Again, be sure to consider the increased complexity of decision-making and comparisons supported by a simple Oneway ANOVA. As an advance organizer, even more complex comparisons and analyses are supported in Twoway ANOVA, but that is not the purpose of this lesson.

The dataset is fairly large (N > 4,000) and there are missing data. Further, the dataset includes both factor-type variables and numeric-type variables that are not used in this lesson. Even though these variables are not used, they need to be accommodated, in part to be sure that the entire dataset is correct.

Be sure to note how numerical codes have been purposely used for some factor-type data in this lesson. Numerical codes are quite common and as such, the creation of a later Code Book is essential so that there is agreement on what each code represents.

5.1.2 Null Hypothesis (Ho)

There is no statistically significant difference ($p <= 0.05$) in M1 values by F2b breakout groups (Group F2b-1, Group F2b-2, Group F2b-3, Group F2b-4, and Group F2b-5).

5.2 Data Import of a .csv Spreadsheet-Type Data File into R

For this lesson, the dataset has been prepared in .csv (e.g., comma-separated values) file format. The data are separated by commas. The data are not separated by tabs and the data are not separated by spaces.

```
####################################################################
# Housekeeping                    Use for All Analyses             #
####################################################################
date()                 # Current system time and date.
R.version.string       # R version and version release date.
ls()                   # List all objects in the working
                       # directory.
rm(list = ls())        # CAUTION: Remove all files in the working
                       # directory. If this action is not desired,
                       # use the rm() function one-by-one to remove
                       # the objects that are not needed.
ls.str()               # List all objects, with finite detail.
getwd()                # Identify the current working directory.
setwd("F:/R_Biostatistics")
                       # Set to a new working directory.
                       # Note the single forward slash and double
                       # quotes.
                       # This new directory should be the directory
                       # where the data file is located, otherwise
                       # the data file will not be found.
getwd()                # Confirm the working directory.
list.files()           # List files at the PC directory.
####################################################################

BioSpmen.df <- read.table (file =
  "Biological_Specimen.csv",
  header = TRUE,
  sep = ",")                       # Import the .csv file

getwd()                            # Identify the working directory
ls()                               # List objects
attach(BioSpmen.df)                # Attach the data, for later use
str(BioSpmen.df)                   # Identify structure
nrow(BioSpmen.df)                  # List the number of rows
ncol(BioSpmen.df)                  # List the number of columns
dim(BioSpmen.df)                   # Dimensions of the data frame
names(BioSpmen.df)                 # Identify names
colnames(BioSpmen.df)              # Show column names
rownames(BioSpmen.df)              # Show row names
head(BioSpmen.df)                  # Show the head
tail(BioSpmen.df)                  # Show the tail
BioSpmen.df                        # Show the entire data frame
summary(BioSpmen.df)               # Summary statistics
```

By completing this action, an object called BioSpmen.df has been created. This R-based object is a data frame and it consists of the data originally included in the file Biological_Specimen.csv, a comma-separated .csv file. To avoid possible conflicts, make sure that there are no prior R-based objects called BioSpmen.df. The prior use of rm(list = ls()) accommodates this concern, removing all prior objects in the current R session.

5.3 Organize the Data and Display the Code Book

Note how it was only necessary to key the filename for the .csv file and not the full pathname since the R working directory is currently set to the directory and/or subdirectory where this .csv file is located (see the Housekeeping section at the beginning of this lesson).

5.3 Organize the Data and Display the Code Book

Now that the data have been imported into R, it is usually necessary to check the data for format and then make any changes that may be needed, to organize the data. As a typical example, consider the common practice of numeric codes, as factors, for Gender. If Gender were coded as 1 and 2 instead of Female (1) and Male (2), it would be necessary to do something so that 1 and 2 are seen as factor (e.g., group) values and not integers more suited for math operations. Of course, this concept applies to all other cases where numeric codes are used to identify factors (e.g., groups). In this lesson, that concept applies to the numeric codes used to identify the five breakout groups for object variables F2b.

For this lesson, the class() function, str() function, and duplicated() function will be used to be sure that data are organized as desired.

```
class(BioSpmen.df)
class(BioSpmen.df$ID)      # DataFrame$ObjectName notation
class(BioSpmen.df$M1)      # DataFrame$ObjectName notation
class(BioSpmen.df$M2)      # DataFrame$ObjectName notation
class(BioSpmen.df$M3a)     # DataFrame$ObjectName notation
class(BioSpmen.df$M3b)     # DataFrame$ObjectName notation
class(BioSpmen.df$M3c)     # DataFrame$ObjectName notation
class(BioSpmen.df$F1)      # DataFrame$ObjectName notation
class(BioSpmen.df$F2a)     # DataFrame$ObjectName notation
class(BioSpmen.df$F2b)     # DataFrame$ObjectName notation

str(BioSpmen.df)           # Structure

duplicated(BioSpmen.df)    # Duplicates
```

The class for each object seems to be correct and there are no duplicate rows of data in the data frame. Saying this, a Code Book will help with future understanding this dataset.

A Code Book is an essential aid for anyone involved in the day-to-day activities of the research and statistics process. The Code Book is typically brief and only serves as a useful reminder for what can be easily forgotten months (or even weeks) later, to make it easy to decipher what may otherwise be seen as arcane numeric codes. Coding schemes that are intuitively obvious today can easily become obtuse tomorrow.

Now that the class(), str(), and duplicated() functions have been used for basic diagnostics, consult the Code Book and coerce each object, as needed, into its correct class.

```
###########################################################
# Code Book                                               #
###########################################################
#                                                         #
# ID                ................... Factor (e.g., nominal) #
#                    A unique ID ranging from 1 to 4,660  #
#                                                         #
# M1 ........................ Numeric (e.g., interval)   #
#                     An unidentified biological variable #
#                         that ranges from 0.00 to 600.0  #
#                                                         #
# M2 ........................ Numeric (e.g., interval)   #
#                     An unidentified biological variable #
#                         that ranges from 0.00 to 10.00  #
#                                                         #
# M3a                         Numeric (e.g., interval)   #
#                     An unidentified biological variable #
#                         that ranges from 0.00 to 4.00   #
#                                                         #
# M3b                         Numeric (e.g., interval)   #
#                     An unidentified biological variable #
#                         that ranges from 0.00 to 4.00   #
#                                                         #
# M3c                         Numeric (e.g., interval)   #
#                     An unidentified biological variable #
#                         that ranges from 0.00 to 4.00   #
#                                                         #
# F1 .........................Factor (e.g., nominal)     #
#                                           Group F1-1   #
#                                           Group F1-2   #
#                                                         #
# F2a........................Factor (e.g., nominal)      #
#                                           Group F2a-1  #
#                                           Group F2a-2  #
#                                                         #
# F2b........................Factor (e.g., nominal)      #
#                                           Group F2b-1  #
#                                           Group F2b-2  #
#                                           Group F2b-3  #
#                                           Group F2b-4  #
#                                           Group F2b-5  #
###########################################################
```

In an effort to promote self-documentation and readability, it is often desirable to label all object variables. First, use the epicalc::des() function and the str() function

5.3 Organize the Data and Display the Code Book

to see the nature of the data frame. Then, use the epicalc::label.var() function to provide descriptive labels for each variable. Of course, be sure to load the epicalc package, if it is not operational from prior analyses.

```
install.packages("epicalc")
library(epicalc)            # Load the epicalc package.
help(package=epicalc)       # Show the information page.
sessionInfo()               # Confirm all attached packages.

epicalc::des(BioSpmen.df)
str(BioSpmen.df)

epicalc::label.var(ID,  "Subject ID",   dataFrame=BioSpmen.df)
epicalc::label.var(M1,  "Variable M1",  dataFrame=BioSpmen.df)
epicalc::label.var(M2,  "Variable M2",  dataFrame=BioSpmen.df)
epicalc::label.var(M3a, "Variable M3a", dataFrame=BioSpmen.df)
epicalc::label.var(M3b, "Variable M3b", dataFrame=BioSpmen.df)
epicalc::label.var(M3c, "Variable M3c", dataFrame=BioSpmen.df)
epicalc::label.var(F1,  "Variable F1",  dataFrame=BioSpmen.df)
epicalc::label.var(F2a, "Variable F2a", dataFrame=BioSpmen.df)
epicalc::label.var(F2b, "Variable F2b", dataFrame=BioSpmen.df)
```

Then confirm the description of each object variable, to be sure that all actions were deployed correctly.

```
epicalc::des(BioSpmen.df)
str(BioSpmen.df)
```

With assurance that the data frame is in correct format and that labels are correct, coerce objects into correct format.

```
BioSpmen.df$ID   <- as.factor(BioSpmen.df$ID)
BioSpmen.df$M1   <- as.numeric(BioSpmen.df$M1)
BioSpmen.df$M2   <- as.numeric(BioSpmen.df$M2)
BioSpmen.df$M3a  <- as.numeric(BioSpmen.df$M3a)
BioSpmen.df$M3b  <- as.numeric(BioSpmen.df$M3b)
BioSpmen.df$M3c  <- as.numeric(BioSpmen.df$M3c)
BioSpmen.df$F1   <- as.factor(BioSpmen.df$F1)
BioSpmen.df$F2a  <- as.factor(BioSpmen.df$F2a)
BioSpmen.df$F2b  <- as.factor(BioSpmen.df$F2b)
```

As a sidebar comment, at the R prompt, key help(as.numeric) and then key help(as.integer) to see the differences between these two R functions and when it may be best to use each.

Again, confirm the structure of the dataset, using both the epicalc::des() function and the str() function.

```
epicalc::des(BioSpmen.df)
str(BioSpmen.df)
```

Then, in a somewhat redundant fashion and to merely further confirm the nature of the dataset, use the base package (obtained when R is first downloaded) levels() function against the factor object variables, to reinforce understanding of the data.

```
levels(BioSpmen.df$F1)
levels(BioSpmen.df$F2a)
levels(BioSpmen.df$F2b)
```

Use the summary() function against the object BioSpmen.df, which is a data frame, to gain an initial sense of descriptive statistics and frequency distributions.

```
summary(BioSpmen.df)
```

Although the dataset seems to be in correct format, it is somewhat difficult to work with numeric values for factor object variables: F1, F2a, and F2b. Use the Code Book to review the meaning for each factor code and then note how this problem is easy to accommodate, but also remember that there are other ways to use R to achieve this aim.

```
BioSpmen.df$F1.recode  <- factor(BioSpmen.df$F1,
   labels = c("Group F1-1",
              "Group F1-2"))    # Apply the labels() function.
head(BioSpmen.df$F1.recode)     # View the first lines of data.
summary(BioSpmen.df$F1.recode)  # View descriptive statistics.

par(ask=TRUE)
epicalc::tab1(BioSpmen.df$F1.recode,  # Use the tab1() function
   decimal=2,                         # from the epicalc
   sort.group=FALSE,                  # package to see details
   cum.percent=TRUE,                  # about the selected
   graph=TRUE,                        # object variable. (The
   missing=TRUE,                      # 1 of tab1 is the one
   bar.values=c("frequency"),         # numeric character and
   horiz=FALSE,                       # it is not the letter l.
   cex=1.15, cex.names=1.15, cex.lab=1.15, cex.axis=1.15,
   main="Factor Levels for Object Variable F1",
   ylab="Frequency of F1 Factor Levels",
   col=c("black", "red", "green"), gen=TRUE)
```

Note how the epicalc::tab1() function is quite useful in that it generates a text-based frequency distribution table and it also generates a frequency distribution graphic. The text-based frequency table is especially helpful in the way it provides cumulative frequencies that account for missing data.

```
BioSpmen.df$F2a.recode  <- factor(BioSpmen.df$F2a,
   labels = c("Group F2a-1",
              "Group F2a-2"))    # Apply the labels() function.
head(BioSpmen.df$F2a.recode)     # View the first lines of data.
summary(BioSpmen.df$F2a.recode)  # View descriptive statistics.
```

5.3 Organize the Data and Display the Code Book

```
par(ask=TRUE)
epicalc::tab1(BioSpmen.df$F2a.recode, # Use the tab1() function
  decimal=2,                          # from the epicalc
  sort.group=FALSE,                   # package to see details
  cum.percent=TRUE,                   # about the selected
  graph=TRUE,                         # object variable. (The
  missing=TRUE,                       # 1 of tab1 is the one
  bar.values=c("frequency"),          # numeric character and
  horiz=FALSE,                        # it is not the letter l.
  cex=1.15, cex.names=1.15, cex.lab=1.15, cex.axis=1.15,
  main="Factor Levels for Object Variable F2a",
  ylab="Frequency of F2a Factor Levels",
  col=c("black", "red", "green"), gen=TRUE)

BioSpmen.df$F2b.recode <- factor(BioSpmen.df$F2b,
  labels = c("Group F2b-1", "Group F2b-2",
             "Group F2b-3", "Group F2b-4",
             "Group F2b-5"))    # Apply the labels() function.
head(BioSpmen.df$F2b.recode)    # View the first lines of data.
summary(BioSpmen.df$F2b.recode) # View descriptive statistics.

par(ask=TRUE)
epicalc::tab1(BioSpmen.df$F2b.recode, # Use the tab1() function
  decimal=2,                          # from the epicalc
  sort.group=FALSE,                   # package to see details
  cum.percent=TRUE,                   # about the selected
  graph=TRUE,                         # object variable. (The
  missing=TRUE,                       # 1 of tab1 is the one
  bar.values=c("frequency"),          # numeric character and
  horiz=FALSE,                        # it is not the letter l.
  cex=1.15, cex.names=1.15, cex.lab=1.15, cex.axis=1.15,
  main="Factor Levels for Object Variable F2b",
  ylab="Frequency of F2b Factor Levels",
  col=c("black", "red", "green", "blue",
        "purple", "sienna"), gen=TRUE)
```

Review the output, below, for the frequency distribution of the object variable F2b.recode. Think about how it would be fairly easy to copy this output and paste it into a word processing program.

```
BioSpmen.df$F2b.recode :
            Frequency  %(NA+)  cum.%(NA+)   %(NA-)  cum.%(NA-)
Group F2b-1      2104   45.15       45.15    45.45       45.45
Group F2b-2       670   14.38       59.53    14.47       59.93
Group F2b-3       707   15.17       74.70    15.27       75.20
Group F2b-4       609   13.07       87.77    13.16       88.36
Group F2b-5       539   11.57       99.33    11.64      100.00
NA's               31    0.67      100.00     0.00      100.00
   Total        4660  100.00      100.00   100.00      100.00
```

Again, there are possibly many ways to recode the factor levels in object variables F1, F2a, and F2b. The method used in this example was fairly simple and used the labels() function.

Now, merely use the attach() function again to assure that all data are attached to the data frame.

```
attach(BioSpmen.df)
head(BioSpmen.df)
tail(BioSpmen.df)
summary(BioSpmen.df)    # Quality assurance data check.

str(BioSpmen.df)        # List all objects, with finite detail.
ls.str(BioSpmen.df)     # List all objects, with finite detail.
```

As an additional data check, use the table() function to see how data have been summarized using the newly created names in the original and coerced factor-type object variables.

```
table(BioSpmen.df$F1,          useNA = c("always"))
table(BioSpmen.df$F1.recode,   useNA = c("always"))

table(BioSpmen.df$F2a,         useNA = c("always"))
table(BioSpmen.df$F2a.recode,  useNA = c("always"))

table(BioSpmen.df$F2b,         useNA = c("always"))
table(BioSpmen.df$F2b.recode,  useNA = c("always"))
```

Note how the argument useNA = c("always") is used with the table function, to force identification of missing values.

This type of redundancy and attention to detail at this stage of development may seem unnecessary, but it more than helps reduce later errors caused by a simple oversight.

5.4 Conduct a Visual Data Check

With the data in proper format, it would be common to immediately attempt the appropriate inferential analyses, Oneway ANOVA for this lesson. However, it is best to first prepare a few graphical displays of the data and to then reinforce comprehension of the data with descriptive statistics and measures of central tendency.

The summary() function, min() function, and max() function are all certainly useful for data checking, but there are also many advantages to a visual data check process. In this case, simple plots can be very helpful in looking for data that may be either illogical or out-of-range. These initial plots will be, by design, simple and should be considered throwaways as they are intended only for initial

5.4 Conduct a Visual Data Check

diagnostic purposes. More complex figures, often of publishable quality can then be prepared from these initial throwaway graphics, by careful selection of functions and arguments.

Although the emphasis in this lesson is on Oneway ANOVA for the factor-type object variable F2b (five breakout groups) and the numeric-type object variable M1 (values can range from 0.00 to 600.00), a simple graphic will be prepared for each variable, largely as a quality assurance check against the entire dataset. Experienced researchers practice quality assurance in multiple ways and at multiple times.

```
names(BioSpmen.df)      # Confirm all object variables.

par(ask=TRUE)
plot(BioSpmen.df$ID,
  main="BioSpmen.df$ID Visual Data Check")

par(ask=TRUE)
plot(density(BioSpmen.df$M1,
  na.rm=TRUE),   # Required for the density() function
  main="Density Plot of M1",
  lwd=6, col="red", font.axis=2, font.lab=2)

par(ask=TRUE)
plot(density(BioSpmen.df$M2,
  na.rm=TRUE),   # Required for the density() function
  main="Density Plot of M2",
  lwd=6, col="red", font.axis=2, font.lab=2)

par(ask=TRUE)
plot(density(BioSpmen.df$M3a,
  na.rm=TRUE),   # Required for the density() function
  main="Density Plot of M3a",
  lwd=6, col="red", font.axis=2, font.lab=2)

par(ask=TRUE)
plot(density(BioSpmen.df$M3b,
  na.rm=TRUE),   # Required for the density() function
  main="Density Plot of M3b",
  lwd=6, col="red", font.axis=2, font.lab=2)

par(ask=TRUE)
plot(density(BioSpmen.df$M3c,
  na.rm=TRUE),   # Required for the density() function
  main="Density Plot of M3c",
  lwd=6, col="red", font.axis=2, font.lab=2)

par(ask=TRUE)
plot(BioSpmen.df$F1.recode,
  main="BioSpmen.df$F1.recode Visual Data Check")
```

```
par(ask=TRUE)
plot(BioSpmen.df$F2a.recode,
  main="BioSpmen.df$F2a.recode Visual Data Check")

par(ask=TRUE)
plot(BioSpmen.df$F2b.recode,
  main="BioSpmen.df$F2b.recode Visual Data Check",
  col=gray(4:0/4))
  # Note the gray shading scheme and how the values range
  # from 0 to 4 and not 1 to 5.  Counts that begin with 0
  # represent a common means of counting in computer science.
```

The purpose of these initial plots is to gain a general sense of the data and to equally look for outliers. In an attempt to look for outliers, the ylim argument has been avoided, so that all data are plotted. Extreme values may or may not be outliers, but they are certainly interesting and demand attention.

This sample lesson has been designed to look into the nature of the numeric-type object variable M1 and the factor-type object variable F2b. Given the nature of M1 values, it may also be a good idea to supplement the plot(density()) function with the hist() function and the boxplot() function, to gain a another view of the continuous values for this object variable. Although object variable M1 does not show perfect normal distribution along a bell-shaped curve, it is assumed that the distribution of M1 approximates those conditions needed for correct use of Oneway ANOVA. Then other functions used in the lattice package and the sm package may have potential use, to further explain how data are organized.

```
par(ask=TRUE)
hist(BioSpmen.df$M1,
  main="BioSpmen.df$M1 Visual Data Check (Histogram)",
  font=2,            # Bold text
  cex.lab=1.15,      # Large font
  col="red")         # Vibrant color

par(ask=TRUE)
boxplot(BioSpmen.df$M1,
  horizontal=TRUE,
  main="Horizontal Boxplot of M1 Values",
  xlab="M1 Values (Limit = 0.00 to 600.00)",
  ylim=c(300,625),       # Note the selection for ylim.
  cex.lab=1.15, cex.axis=1.15, border="blue", col="red")
box()

par(ask=TRUE)
boxplot(BioSpmen.df$M1 ~ BioSpmen.df$F2b.recode,
  horizontal=FALSE,
  main="Vertical Boxplot of M1 Values by F2b Groups",
  ylim=c(350,550), # Adjust range for mix() and max()
```

5.4 Conduct a Visual Data Check

```
   ylab="M1 Values (Limit = 0.00 to 600.00)",
   xlab="F2b Groups",
   cex.lab=1.15, cex.axis=1.15, border="blue", col="red")
box()
```

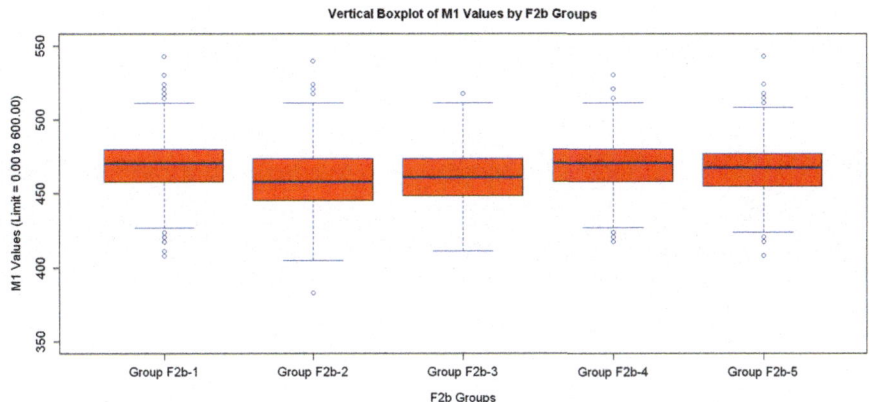

If all group names do not appear in the graphic, adjust the cex-type settings to a lower value. As is nearly always the case with R, settings are generally a matter of balance between personal preferences and presentation requirements. And remember that these settings can always be reviewed by keying help(function.name), to learn more about the R function and the many arguments and options supported by the function, such as help(par).

Note: It is largely a personal preference to display a boxplot in either horizontal mode or vertical mode. To meet the needs of various readers, it is common to use both orientations.

```
install.packages("lattice")
library(lattice)              # Load the lattice package.
help(package=lattice)         # Show the information page.
sessionInfo()                 # Confirm all attached packages.

par(ask=TRUE) # 1 Column by 1 Row Histogram
lattice::histogram(~ BioSpmen.df$M1,
   type="count", # Note: count
   par.settings=simpleTheme(lwd=2),
   par.strip.text=list(cex=1.15, font=2),
   scales=list(cex=1.15),
   main="Histogram (Count) of M1",
   xlab=list("M1", cex=1.15, font=2),
   xlim=c(300,600), # Note the range.
   ylab=list("Count", cex=1.15, font=2),
   aspect=1, breaks=10,
   layout = c(1,1), # Note: 1 Column by 1 Row.
   col="red")
```

```
par(ask=TRUE) # 1 Column by 5 Rows Histogram
lattice::histogram(~ BioSpmen.df$M1 |
  BioSpmen.df$F2b.recode, # Pipe, not ~ or ,
  type="count", # Note: count
  par.settings=simpleTheme(lwd=2),
  par.strip.text=list(cex=1.15, font=2),
  scales=list(cex=1.15),
  main="Histograms (Count) of M1 by F2b.recode",
  xlab=list("M1", cex=1.15, font=2),
  xlim=c(300,600), # Note the range.
  ylab=list("Count", cex=1.15, font=2),
  aspect=0.25, breaks=10,
  layout = c(1,5), # Note: 1 Column by 5 Rows.
  col="red")

par(ask=TRUE) # 1 Column by 5 Rows Histogram
lattice::histogram(~ BioSpmen.df$M1 |
  BioSpmen.df$F2b.recode, # Pipe, not ~ or ,
  type="density", # Note: density
  par.settings=simpleTheme(lwd=2),
  par.strip.text=list(cex=1.15, font=2),
  scales=list(cex=1.15),
  main="Histograms (Density) of M1 by F2b.recode",
  xlab=list("M1",
  cex=1.15, font=2),
  xlim=c(300,600), # Note the range.
  ylab=list("Density", cex=1.15, font=2),
  aspect=0.25, breaks=10,
  layout = c(1,5), # Note: 1 Column by 5 Rows.
  col="red")

par(ask=TRUE) # Breakout group by measured object.
lattice::bwplot(BioSpmen.df$F2b.recode ~
  BioSpmen.df$M1,           # Tilde, not | or ,
  par.settings = simpleTheme(lwd=2),
  par.strip.text=list(cex=1.15, font=2),
  scales=list(cex=1.15),
  main="Boxplot of M1 by F2b Breakout Groups",
  xlab=list("M1", cex=1.15, font=2),
  xlim=c(300,600),
  ylab=list("F2b Breakout Groups", cex=1.15, font=2),
  aspect=0.5, layout=c(1,1), col="red")

install.packages("sm")
library(sm)                 # Load the sm package.
help(package=sm)            # Show the information page.
sessionInfo()               # Confirm all attached packages.

saveline.width <- par(lwd=6)          # Generate a heavy line
```

```
savecex.axis   <- par(cex.axis=1.25)   # Adjust axis
par(ask=TRUE)
sm::sm.density.compare(BioSpmen.df$M1, BioSpmen.df$F2b.recode,
  xlab=list("M1 Values (Limit = 0.00 to 600.00)",
  cex=1.15, font=2),
  ylab=list("Density", cex=1.25, font=2),
  xlim=c(300,650), # Adjust to actual values
  ylim=c(0,0.03))  # Experiment with the ylim values
title(main="Density Plot of M1 Values by F2b Groups")
colorfill <- c(2:(2+length(levels(BioSpmen.df$F2b.recode))))
legend(locator(1), levels(BioSpmen.df$F2b.recode),
  fill=colorfill)
par(saveline.width); par(savecex.axis)
  # Note how the line width is accommodated and then set back
  # to the original value.
  # Place the legend in any desired location by clicking the
  # left mouse button.
```

5.5 Descriptive Analysis of the Data

This dataset continues the need for attention to missing data. Given the different ways missing data can impact analyses, it is often helpful to first check for missing data by using the is.na() function and the complete.cases() function against the entire dataset. Both functions return a TRUE or FALSE response, depending on the function and the outcome of whether data are missing or data are not missing.

```
is.na(BioSpmen.df)            # Check for missing data
complete.cases(BioSpmen.df)   # Check for complete cases
```

For the dataset BioSpmen.df note how there are many rows where there are missing data for individual object variables. The exact nature of the dataset is not known and for now there is no judgment placed on why there are missing data. For the purpose of this lesson, it is only necessary to recognize that there are missing data and to use appropriate functions and arguments to accommodate this observation.

Even for this fairly large dataset, the summary() function may be all that is necessary to gain a sense of the data. Note how the summary() function is applied against the entire dataset, thus yielding information about all object variables, including the object variable ID.

```
summary(BioSpmen.df)
```

Give attention to the listing of NAs for those object variables with missing data. Again, the summary() function is very useful and it should always be a first selection when preparing descriptive analyses.

Although the summary() function is quite sufficient, descriptive statistics for individual object variables may be desired. To achieve this aim, review the prior lesson *Descriptive Statistics and Measures of Central Tendency* for a comprehensive review of the functions used for descriptive statistics, especially: length(), asbio::Mode(), median(), mean(), sd(), table(), and finally summary(). As needed (but not always, depending on specific functions), the na.rm=TRUE argument or some other similar convention will be used to accommodate missing data.

```
length(BioSpmen.df$F2b)                    # N of F2b
length(BioSpmen.df$M1)                     # N of M1

mode.of.M1 <-
  names(sort(-table(BioSpmen.df$M1)))[1]
mode.of.M1      # Hand calculate mode with NAs

median(BioSpmen.df$M1, na.rm=TRUE)          # Median

mean(BioSpmen.df$M1, na.rm=TRUE)            # Mean
sd(BioSpmen.df$M1,na.rm=TRUE )              # SD
   # Measures of Central Tendency

summary(BioSpmen.df)
```

Descriptive statistics at the summary level are always useful, but breakout statistics are also needed to gain a more complete understanding of the data. There are many ways to obtain breakout statistics, but the tapply() function, epicalc::summ() function, prettyR::brkdn() function, psych:::describe.by() function, Hmisc::bystats() function, and the lessR::SummaryStats() function are among the most detailed and easiest to use, to discern differences between breakout groups such as the breakout groups for BioSpmen.df$F2b.recode associated with this lesson: Group F2b-1, Group F2b-2, Group F2b-3, Group F2b-4, and Group F2b-5.

```
tapply(M1, F2b.recode, summary, na.rm=TRUE, data=BioSpmen.df)
   # M1 by F2b.recode, using tapply()

par(ask=TRUE) # Use the epicalc package for breakout analyses
epicalc::summ(BioSpmen.df$M1, by=BioSpmen.df$F2b.recode,
   graph=TRUE, pch=18, ylab="auto",
   main="Sorted Dotplot of M1 by F2b.recode",
   cex.X.axis=1.15, cex.Y.axis=1.15, font.lab=2, dot.col="auto")
   # Note the descriptive statistics and not only the graphic
   # that go along with the  epicalc::summ() function.

install.packages("prettyR")
library(prettyR)                  # Load the prettyR package.
help(package=prettyR)             # Show the information page.
sessionInfo()                     # Confirm all attached packages.
```

5.6 Conduct the Statistical Analysis

```
prettyR::describe.factor(BioSpmen.df$F2b.recode)
prettyR::describe.numeric(BioSpmen.df$M1)
prettyR::brkdn(M1 ~ F2b.recode, BioSpmen.df)

install.packages("psych")
library(psych)              # Load the psych package.
help(package=psych)         # Show the information page.
sessionInfo()               # Confirm all attached packages.

psych::describe.by(BioSpmen.df$M1, BioSpmen.df$F2b.recode,
  mat=TRUE)                 # Matrix output

install.packages("Hmisc")
library(Hmisc)              # Load the Hmisc package.
help(package=Hmisc)         # Show the information page.
sessionInfo()               # Confirm all attached packages.

Hmisc::bystats(BioSpmen.df$M1, BioSpmen.df$F2b.recode,
  nmiss=TRUE)

 Mean of BioSpmen.df$M1 by

                N  Missing     Mean
Group F2b-1  2064       40 469.8119
Group F2b-2   650       20 460.5897
Group F2b-3   691       16 461.4755
Group F2b-4   600        9 469.9272
Group F2b-5   515       24 466.9942
NA              9       22 457.0444
ALL          4529      131 466.8859

install.packages("lessR")
library(lessR)              # Load the lessR package.
help(package=lessR)         # Show the information page.
sessionInfo()               # Confirm all attached packages.

lessR::SummaryStats(M1, by=F2b.recode, data=BioSpmen.df,
  digits.d=2)  # Force to two places beyond the decimal
```

5.6 Conduct the Statistical Analysis

The preceding graphical images and descriptive statistics, both summary descriptive statistics and breakout descriptive statistics, provide a fairly good idea of the M1 values, overall and by breakout group (F2b.recode breakout groups in this lesson):

```
Breakout Descriptive Statistics of M1 Values by F2b.recode in
Ascending Rank Order by Mean
```

```
          group      n    mean      sd median
Group F2b-2    650  460.59  19.729 458.44
Group F2b-3    691  461.48  17.798 461.58
Group F2b-5    515  466.99  18.198 467.86
Group F2b-1   2064  469.81  16.981 471.00
Group F2b-4    600  469.93  17.498 471.00
```

5.6.1 Exploratory Oneway ANOVA

The prior graphical images and descriptive statistics are all certainly useful and provide a sense of the data. Even so, for this lesson a Oneway ANOVA will help determine if mean differences by breakout groups (e.g., Group F2b-1, Group F2b-2, Group F2b-3, Group F2b-4, Group F2b-5) for M1 are true differences at p <= 0.05 or if the differences are instead due only to chance.

Perhaps the simplest Oneway ANOVA test calls for use of the oneway.test() function. View the use of the oneway.test() function as an exploratory tool for Oneway ANOVA. The output is fairly terse, but sufficient information is provided to determine if there is a statistically significant difference in means by breakout groups, or means for M1 by F2b.recode breakout groups in this lesson. However, little else is provided.

```
oneway.test(M1 ~ F2b.recode, data = BioSpmen.df,
  var.equal=FALSE)  # Assume variance is not equal

# Outcome:   F = 51.914, p-value < 0.00000000000000022

oneway.test(M1 ~ F2b.recode, data = BioSpmen.df,
  var.equal=TRUE)   # Assume variance is equal

# Outcome:   F = 55.036, p-value < 0.00000000000000022
```

Consistently, both applications of the oneway.test() function provides evidence that there is a statistically significant difference in M1 values by F2b.recode breakout groups. That is to say, calculated p is <= 0.00000000000000022 which is certainly less than the criterion p value of 0.05.

However, other Oneway ANOVA functions are available with the base set of R functions and these other functions are far more informative than output from the simple oneway.test() function. The desire with these other functions is to obtain detailed analyses to ultimately determine not only if there are differences in mean values for M1 by F2b.recode breakout groups, but exactly which breakout groups

5.6 Conduct the Statistical Analysis

show similarity and which breakout groups show differences. To achieve these aims, the lm() function, anova() function, aov() function, and the TukeyHSD() function will all be used to determine if there are differences in M1 values by F2b.recode breakout groups and where those differences exist.

5.6.2 Oneway ANOVA Method 1: lm() and anova() Functions

There are more than a few ways to accommodate a Oneway ANOVA of M1 values by F2b subgroups. The first method demonstrated below calls for:

- Create a new object that is the output of applying the lm() function against the identified data, to fit a linear model.
- Apply the anova() function against this new object, to generate an Analysis of Variance table, including: Degrees of Freedom (Df), Sum of Squares (Sum Sq), Mean Square (Mean Sq), F value, and Calculated p-value (Pr(>F).
- Apply the summary() function against this new object, to gain a sense of output and the overall level of significance (p-value).

```
M1.by.F2b.ANOVA.Method1 <-   lm(M1 ~ F2b.recode -1,
   data=BioSpmen.df, na.action=na.exclude)
   # Create the object M1.by.F2b.ANOVA.Method1

M1.by.F2b.ANOVA.Method1

anova(M1.by.F2b.ANOVA.Method1)

Analysis of Variance Table

Response: M1
              Df    F value                    Pr(>F)
F2b.recode    5     626809   < 0.00000000000000022
Residuals    4515

summary(M1.by.F2b.ANOVA.Method1)
```

Note how this analysis accommodates missing data. The -1 argument provides a fit model without an intercept. Groups are compared only to each other.

This method, based on use of the lm() function, provides evidence that there is a statistically significant difference at p <= 0.05 for M1 values and F2b.recode breakout groups. Note, however, that groupwise comparisons are absent. It is currently unclear which F2b subgroups differ in statistical significance (p <= 0.05) of M1 values.

5.6.3 Oneway ANOVA Method 2: aov() and TukeyHSD() Functions

For greater specificity in determining groupwise differences for the measured object variable, it is necessary to use some type of group mean comparison technique. The common TukeyHSD() function is used in this lesson to provide that level of detail in the examination of M1 values by F2b.recode breakout groups. The Oneway ANOVA method demonstrated below calls for:

- Create an object that is the result of the aov() function applied against the measured object variable and the grouping object variable.
- Apply the summary() function against this new object, to gain a sense of ANOVA table output and the overall level of calculated significance (p-value).
- Apply the TukeyHSD() function against this new object to view group member by group member p-value comparisons, to see which group members have similar outcomes for the measured object variable and which group members have different outcomes for the measured object variable, within the context of p-values. As a reminder of different mean comparison techniques, the TukeyHSD() function refers to Tukey's Honest Significant Difference.

If a significant difference is found, give attention to how the standard nomenclature of ***, **, and * is used to identify significance levels.

Given this background, note how the aov() function in tandem with the TukeyHSD() function are more than useful in an attempt to understand outcomes from a Oneway ANOVA.

```
M1.by.F2b.ANOVA.Method2 <- aov(M1 ~ F2b.recode,
  data=BioSpmen.df)
  # Create the object M1.by.F2b.ANOVA.Method2

M1.by.F2b.ANOVA.Method2

summary(M1.by.F2b.ANOVA.Method2)

TukeyHSD(M1.by.F2b.ANOVA.Method2)
  # Multiple comparisons.
  # This analysis accommodates missing data.

  Tukey multiple comparisons of means
    95% family-wise confidence level

Fit: aov(formula = M1 ~ F2b.recode, data = BioSpmen.df)

$F2b.recode
                             p adj
Group F2b-2-Group F2b-1    0.0000000
Group F2b-3-Group F2b-1    0.0000000
```

```
Group F2b-4-Group F2b-1    0.9999132
Group F2b-5-Group F2b-1    0.0110770
Group F2b-3-Group F2b-2    0.8915424
Group F2b-4-Group F2b-2    0.0000000
Group F2b-5-Group F2b-2    0.0000001
Group F2b-4-Group F2b-3    0.0000000
Group F2b-5-Group F2b-3    0.0000010
Group F2b-5-Group F2b-4    0.0467172
```

In this sample, M1.by.F2b.ANOVA.Method2 was the result of the aov() function applied against the measured object variable (M1) and the grouping object variable (F2b.recode).

The summary() function provided evidence that there was an overall statistically significant difference at p <= 0.05. Note how the calculated p-value (p <= 0.0000000000000002) is indeed less than 0.05.

The TukeyHSD() function provided more specificity, giving detailed outcomes on where there are statistically significant differences (p <= 0.05) for M1 values by F2b.recode breakout groups and equally, were there is no difference for M1 values by F2b subgroups.

5.7 Summary

In this lesson, the graphics and statistics provided a great deal of information. Of immediate importance, however, focus on the Null Hypothesis statement and the breakout analyses deriving from use of the TukeyHSD() function:

Null Hypothesis (Ho): There is no statistically significant difference (p <= 0.05) in M1 values by F2b breakout groups (Group F2b-1, Group F2b-2, Group F2b-3, Group F2b-4, and Group F2b-5).

For this lesson, there are overall statistically significant difference (p <= 0.05) in M1 values by F2b.recode breakout groups. Oneway ANOVA served as the primary approach to determine if a statistically significant difference existed and the TukeyHSD() function then served as the tool by which finite comparisons of M1 by F2b.recode comparisons were made

Review the table immediately below, made by editing the output associated with this lesson, to see which F2b.recode breakout groups are significantly different from each other (p <= 0.05) and which are not.

```
Comparison of F2b Breakout Groups to M1 Mean Values and
Statistical Significance (p <= 0.05)

Group            Mean     Group            Mean      p adj       p <= 0.05
Group F2b-2     460.59    Group F2b-1     469.81    0.00000     Yes
Group F2b-3     461.48    Group F2b-1     469.81    0.00000     Yes
Group F2b-4     469.93    Group F2b-1     469.81    0.99991     No
```

```
Group F2b-5   466.99   Group F2b-1   469.81   0.01108   Yes
Group F2b-3   461.48   Group F2b-2   460.59   0.89154   No
Group F2b-4   469.93   Group F2b-2   460.59   0.00000   Yes
Group F2b-5   466.99   Group F2b-2   460.59   0.00000   Yes
Group F2b-4   469.93   Group F2b-3   461.48   0.00000   Yes
Group F2b-5   466.99   Group F2b-3   461.48   0.00000   Yes
Group F2b-5   466.99   Group F2b-4   469.93   0.04672   Yes
```

Overall, there is no statistically significant difference ($p <= 0.05$) in M1 values for the following F2b.recode breakout groups: (1) Group F2b-4 (Mean = 469.93) and Group F2b-1 (Mean = 469.81) and (2) Group F2b-3 (Mean = 461.48) and Group F2b-2 (Mean = 460.59). For all other comparisons, differences are statistically significant ($p <= 0.05$).

Give attention, however, to the calculated p value of 0.04672 for comparison of Group F2b-5 (Mean = 466.99) and Group F2b-4 (Mean = 469.93). As time permits, note how a p value of 0.04672 begins to approximate the declared p value of 0.05 and then read on this distinction between statistical difference and practical difference, but more discussion of this topic is beyond the purpose of this lesson.

A graphical representation of the TukeyHSD values may help better reinforce the group comparisons. To achieve this aim, merely create an object that holds the TukeyHSD values and then plot these values.

```
M1.hsd <- TukeyHSD(M1.by.F2b.ANOVA.Method2)

M1.hsd
```

A plot of M1.hsd is generally helpful for when only a few subgroups are compared to each other. See below if in this lesson, where there are five subgroups, if the plot is of any practical value for presentation to others.

```
savefont     <- par(font=2)      # Font bold
savecex.lab  <- par(cex=0.50)    # Label size
par(ask=TRUE)
plot(M1.hsd, las=3, col="red")
par(savefont) ; par(savecex.lab)
```

Given the complexity of this figure and the need for small font size, other graphical images may better reinforce the outcomes of this Oneway ANOVA, where M1 values have been analyzed for the five F2b.recode breakout groups.

The qplot() function using the ggplot2 package can be used to create a simple graphic, reinforcing distribution of M1 values by F2b.recode breakout groups.

```
install.packages("ggplot2")
library(ggplot2)              # Load the ggplot2 package.
help(package=ggplot2)         # Show the information page.
sessionInfo()                 # Confirm all attached packages.

par(ask=TRUE)
```

5.7 Summary

```
ggplot2::qplot(F2b.recode, M1,
  data=BioSpmen.df,
  main="M1 Values by F2b Subgroups",
  geom = "auto",
  position="dodge")+theme_bw()
# Note the placement of +theme_bw().
```

The ggplot2::qplot() function is interesting, but the stripchart() function is also useful and it may yield a more complete view of how data are distributed, allowing another comparison of M1 values by F2b.recode breakout groups.

```
par(ask=TRUE)
stripchart(BioSpmen.df$M1~BioSpmen.df$F2b.recode,
  method="jitter", jitter=.1, vertical=TRUE,
  main="Jitter Stripchart of M1 Values by F2b Subgroups",
  xlab="F2b Subgroups", ylab="M1 values", cex.lab=1.25,
  cex.axis=1.25, ylim=c(325,600), pch=19, col="darkred")
```

Finally, go back to the graphical images at the level of group comparisons (Mean and Confidence Interval) to see how these images parallel the Oneway ANOVA findings. Syntax for complementary graphical images of M1 by F2b.recode follows, based on use of the sciplot::lineplot() function and the s20x::boxqq() function.

```
install.packages("sciplot")
library(sciplot)           # Load the sciplot package.
help(package=sciplot)      # Show the information page.
sessionInfo()              # Confirm all attached packages.

par(ask=TRUE)
sciplot::lineplot.CI(F2b.recode, M1,
  main="Mean and CI M1 Values by F2b Subgroups:
  M1 Scale = 450 to 475",
  xlab="F2b Subgroup", ylab="M1 Value",
  font=2, font.lab=2, legend=FALSE, type="b",
  x.cont=FALSE, lwd=3, col="red",
  err.width=.5, err.col="darkblue", err.lty="solid",
  ylim=c(450,475))
  # Notice how the Mean and CI of M1 for each F2b.recode
  # breakout group is easier to see by adjusting the M1
  # scale (450 to 475) on the Y axis.

install.packages("s20x")
library(s20x)              # Load the s20x package.
help(package=s20x)         # Show the information page.
sessionInfo()              # Confirm all attached packages.

par(ask=TRUE)
s20x::boxqq(M1 ~ F2b.recode, data=BioSpmen.df)
```

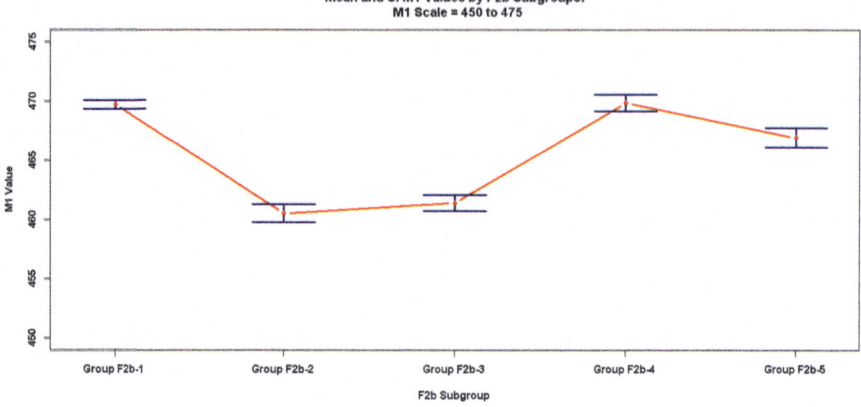

5.8 Addendum: Other Packages for Display of Oneway ANOVA

An advantage of R is that the user community has contributed literally thousands of additional packages that supplement the many functions available in the base package, when R is first downloaded. Not surprisingly, Oneway ANOVA is addressed in some of these additional packages. A few are demonstrated in this addendum, focusing on the functions included in the lessR package and the s20x package.

```
par(ask=TRUE)
lessR::ANOVA(M1 ~ F2b.recode, data=BioSpmen.df,
  digits.d=2, res.rows="all")
```

The lessR::ANOVA() function generates descriptive statistics, an ANOVA table, a Oneway table, Tukey multiple comparisons of means and residuals, and separate graphics.

```
par(ask=TRUE)
s20x::levene.test(M1 ~ factor(F2b.recode),
  show.table=TRUE, BioSpmen.df)
```

```
par(ask=TRUE)
s20x::onewayPlot(M1 ~ factor(F2b.recode),
  BioSpmen.df, strip=FALSE, vert= TRUE,
  verbose=TRUE, conf.level=0.95, pooled=TRUE,
  interval.type="tukey")
```

```
par(ask=TRUE)
s20x::summary1way(lm(M1 ~ F2b.recode,
  BioSpmen.df), inttype="tukey", print.out=TRUE,
```

5.9 Prepare to Exit, Save, and Later Retrieve This R Session

```
raw.plot=TRUE)
# Note how the lm() function was used in this
# presentation.
```

The s20x package has more than a few functions that are related to Analysis of Variance and it should be examined as a supplement to the use of standard functions.

An additional package that may have value is the granova package, which is also used for graphical display of ANOVA-type activities.

Whatever test is selected, be sure to review the text output and not only the figures. A great deal of useful information, beyond what is regularly provided with a standard Oneway ANOVA test, is available through use of various functions and arguments found in external packages.

5.9 Prepare to Exit, Save, and Later Retrieve This R Session

It is common to prepare R syntax in a separate file, using a simple ASCII text editor. If time permits, experiment with Crimson Editor, Tinn-R, or vim, but there are many other possible selections.

```
getwd()              # Identify the current working directory.
ls()                 # List all objects in the working
                     # directory.
ls.str()             # List all objects, with finite detail.
list.files()         # List files at the PC directory.

save.image("R_Lesson_Oneway_ANOVA.rdata")

getwd()              # Identify the current working directory.
ls()                 # List all objects in the working
                     # directory.
ls.str()             # List all objects, with finite detail.
list.files()         # List files at the PC directory.

alarm()              # Alarm, notice of upcoming action.
q()                  # Quit this session.
                     # Prepare for Save workspace image? query.
```

Use the R Graphical User Interface (GUI) to load the saved rdata file: **File** and then **Load Workspace**. Otherwise, use the load() function, keying the full pathname, to load the .rdata file and retrieve the session.

Recall, however, that it may be just as useful to simply use a R script file (typically saved as a .txt ASCII-type file) and recreate the analyses and graphics, provided the data files remain available.

Chapter 6
Twoway Analysis of Variance (ANOVA)

Abstract The purpose of this lesson on Twoway Analysis of Variance (ANOVA) is to provide guidance on how R can be used to see if two or more group means (e.g., Mean of Variable X for Gender breakouts and Race breakouts, Mean of Variable Y for Breed breakouts and Feeding Program Breakouts, etc.) differ due to chance, or if observed differences are indeed the result of true difference between phenomena. Specifically, the Twoway ANOVA statistical test has been structured to examine differences (and possible interactions) when variables have two or more separate categories. The degree of complexity supported by Twoway ANOVA begins to model real-world concerns for the many ways variables, and possible interaction between variables, impacts outcomes.

Keywords ANOVA • Analysis of Variance • Factorial design • Interaction • Mean comparison technique • Oneway ANOVA • Twoway ANOVA

6.1 Background on This Lesson

6.1.1 Description of the Data

This lesson on Twoway Analysis of Variance (ANOVA) focused on human subjects as the biological specimen in question and specifically attempts to determine if there are differences in Systolic Blood Pressure by Gender and by Race. The data are from a sample of 2,000 subjects and there are no missing data.

When Twoway ANOVA is used, this one test makes it is possible to determine:

- Is there a difference because of variables acting independently of each other?
- Is there a difference because of joint effects (i.e., interaction)?

Along with instruction on the use of R and R syntax associated with Twoway ANOVA, this lesson will also reinforce the use of graphical figures and descriptive statistics to complement outcomes from the parametric Twoway ANOVA. A number of external R-based packages and functions are also introduced in this lesson. Numerical codes are not used for the factor-type data in this lesson. The terms Female and Male are used for Gender and the terms Black, Other, and White have been used for Race.

Twoway ANOVA designs, along with all factorial-type ways by which data are organized, can become exceedingly complex. This lesson serves as a first introduction to the use of R with Twoway ANOVA. Separate resources should be viewed on research design, independent of selected statistical analysis software.

Statistical tests that can account for increased complexity are needed if meaningful decisions involving statistics in the large are to be effected. Again, be sure to consider the increased complexity of decision-making and comparisons supported by increasingly complex factorial designs, such as Twoway ANOVA.

6.1.2 Null Hypothesis (Ho)

There is no statistically significant difference (p <= 0.05) in Systolic Blood Pressure by Gender (Female and Male), by Race (Black, Other, White), and by the interaction of Gender and Race.

6.2 Data Import of a .csv Spreadsheet-Type Data File into R

For this lesson, the dataset has been prepared in .csv (e.g., comma-separated values) file format. The data are separated by commas. The data are not separated by tabs and the data are not separated by spaces.

```
################################################################
# Housekeeping                           Use for All Analyses   #
################################################################
date()              # Current system time and date.
R.version.string    # R version and version release date.
ls()                # List all objects in the working
                    # directory.
rm(list = ls())     # CAUTION: Remove all files in the working
                    # directory. If this action is not desired,
                    # use the rm() function one-by-one to remove
                    # the objects that are not needed.
ls.str()            # List all objects, with finite detail.
getwd()             # Identify the current working directory.
setwd("F:/R_Biostatistics")
                    # Set to a new working directory.
                    # Note the single forward slash and double
                    # quotes.
```

6.3 Organize the Data and Display the Code Book

```
                    # This new directory should be the directory
                    # where the data file is located, otherwise
                    # the data file will not be found.
getwd()             # Confirm the working directory.
list.files()        # List files at the PC directory.
###############################################################

Systolic.df <- read.table (file =
  "GenderRaceBPWeight.csv",
  header = TRUE,
  sep = ",")                    # Import the .csv file

getwd()                         # Identify the working directory
ls()                            # List objects
attach(Systolic.df)             # Attach the data, for later use
str(Systolic.df)                # Identify structure
nrow(Systolic.df)               # List the number of rows
ncol(Systolic.df)               # List the number of columns
dim(Systolic.df)                # Dimensions of the data frame
names(Systolic.df)              # Identify names
colnames(Systolic.df)           # Show column names
rownames(Systolic.df)           # Show row names
head(Systolic.df)               # Show the head
tail(Systolic.df)               # Show the tail
Systolic.df                     # Show the entire data frame
summary(Systolic.df)            # Summary statistics
```

By completing these actions, an object called Systolic.df has been created and accommodated. This R-based object is a data frame and it consists of the data originally included in the file GenderRaceBPWeight.csv, a comma-separated .csv file. To avoid possible conflicts, make sure that there are no prior R-based objects called Systolic.df. The prior use of rm(list = ls()) accommodates this concern, removing all prior objects in the current R session.

Note how it was only necessary to key the filename for the .csv file and not the full pathname since the R working directory is currently set to the directory and/or subdirectory where this .csv file is located (see the Housekeeping section at the beginning of this lesson).

6.3 Organize the Data and Display the Code Book

Now that the data have been imported into R, it is usually necessary to check the data for format and then make any changes that may be needed, to organize the data. For this dataset, English-text has been used to distinguish between factor-type object variable breakouts: Gender (Female and Male) and Race (Black, Other, White). Recoding is not needed in this lesson.

For this lesson, the class() function, str() function, and duplicated() function will be used to be sure that data are organized as desired.

```
class(Systolic.df)
class(Systolic.df$ID)          # DataFrame$ObjectName notation
class(Systolic.df$Gender)      # DataFrame$ObjectName notation
class(Systolic.df$Race)        # DataFrame$ObjectName notation
class(Systolic.df$Systolic)    # DataFrame$ObjectName notation
class(Systolic.df$Diastolic)   # DataFrame$ObjectName notation
class(Systolic.df$Weight)      # DataFrame$ObjectName notation

str(Systolic.df)               # Structure
duplicated(Systolic.df)        # Duplicates
```

The class for each object seems to be correct and there are no duplicate rows of data in the data frame. Saying this, a Code Book will help with future understanding this dataset.

A Code Book is an essential aid for anyone involved in the day-to-day activities of the research and statistics process. The Code Book is typically brief and only serves as a useful reminder for what can be easily forgotten months (or even weeks) later, to make it easy to decipher what may otherwise be seen as arcane numeric codes. Coding schemes that are intuitively obvious today can easily become obtuse tomorrow.

Now that the class(), str(), and duplicated() functions have been used for basic diagnostics, consult the Code Book and coerce each object, as needed, into its correct class.

```
############################################################
# Code Book                                                #
############################################################
#                                                          #
# ID ......................... Factor (e.g., nominal)      #
#               A unique ID ranging from S0001 to S2000    #
#                                                          #
# Gender ......................Factor (e.g., nominal)      #
#                                                  Female  #
#                                                    Male  #
#                                                          #
# Race ........................Factor (e.g., nominal)      #
#                                                   Black  #
#                                                   Other  #
#                                                   White  #
#                                                          #
# Systolic ................ Numeric (e.g., interval)       #
#                  Systolic Blood Pressure that ranges     #
#                                  from <= 70 to >= 200    #
#                                                          #
```

6.3 Organize the Data and Display the Code Book

```
# Diastolic ............... Numeric (e.g., interval) #
#              Diastolic Blood Pressure that ranges #
#                           from <= 50 to >= 120    #
#                                                   #
# Weight .................. Numeric (e.g., interval) #
#              Weight (Lbs.) of Adults, ranging     #
#                           from <= 90 to >= 300    #
#####################################################
```

In an effort to promote self-documentation and readability, it is often desirable to label all object variables. First, use the epicalc::des() function and the str() function to see the nature of the data frame. Then, use the epicalc::label.var() function to provide descriptive labels for each variable. Of course, be sure to load the epicalc package, if it is not operational from prior analyses.

```
install.packages("epicalc")
library(epicalc)              # Load the epicalc package.
help(package=epicalc)         # Show the information page.
sessionInfo()                 # Confirm all attached packages.

epicalc::des(Systolic.df)
str(Systolic.df)

epicalc::label.var(ID,        "Subject ID", dataFrame=Systolic.df)
epicalc::label.var(Gender,    "Gender",     dataFrame=Systolic.df)
epicalc::label.var(Race,      "Race",       dataFrame=Systolic.df)
epicalc::label.var(Systolic,  "Systolic",   dataFrame=Systolic.df)
epicalc::label.var(Diastolic, "Diastolic",  dataFrame=Systolic.df)
epicalc::label.var(Weight,    "Weight",     dataFrame=Systolic.df)
```

Then confirm the description of each object variable, to be sure that all actions were deployed correctly.

```
epicalc::des(Systolic.df)
str(Systolic.df)
```

With assurance that the data frame is in correct format and that labels are correct, coerce objects into correct format. Notice how variables are named: DataFrame$ObjectName.

```
Systolic.df$ID        <- as.factor(Systolic.df$ID)
Systolic.df$Gender    <- as.factor(Systolic.df$Gender)
Systolic.df$Race      <- as.factor(Systolic.df$Race)
Systolic.df$Systolic  <- as.numeric(Systolic.df$Systolic)
Systolic.df$Diastolic <- as.numeric(Systolic.df$Diastolic)
Systolic.df$Weight    <- as.numeric(Systolic.df$Weight)
```

As a sidebar comment, at the R prompt, key help(as.numeric) and then key help(as.integer) to see the differences between these two R functions and when it may be best to use each.

Again, confirm the structure of the dataset, using both the epicalc::des() function and the str() function.

```
epicalc::des(Systolic.df)
str(Systolic.df)
```

Then, in a somewhat redundant fashion and to merely further confirm the nature of the dataset, use the base package (obtained when R is first downloaded) levels() function against the factor object variables, to reinforce understanding of the data.

```
levels(Systolic.df$Gender)
levels(Systolic.df$Race)
```

Use the summary() function against the object Systolic.df, which is a data frame, to gain an initial sense of descriptive statistics and frequency distributions.

```
summary(Systolic.df)
```

Now, merely use the attach() function again to assure that all data are attached to the data frame.

```
attach(Systolic.df)
head(Systolic.df)
tail(Systolic.df)
summary(Systolic.df)    # Quality assurance data check

str(Systolic.df)        # List all objects, with finite detail
ls.str(Systolic.df)     # List all objects, with finite detail
```

This type of redundancy and attention to detail at this stage of development may seem unnecessary, but it more than helps reduce later errors caused by a simple oversight.

6.4 Conduct a Visual Data Check

Now that the data are all in proper format, it would be common to immediately attempt the appropriate inferential analyses, Twoway ANOVA for this lesson. However, it is best to first prepare a few graphical displays of the data and to then reinforce comprehension of the data with descriptive statistics and measures of central tendency.

Although the emphasis in this lesson is on Twoway ANOVA for the factor-type object variables Gender (two breakout groups), Race (three breakout groups) and the numeric-type object variable Systolic (values can range from <= 70 to >= 200), a simple graphic will be prepared for each variable, largely as a quality assurance check against the entire dataset.

6.4 Conduct a Visual Data Check

```
names(Systolic.df)        # Confirm all object variables.

par(ask=TRUE)
plot(Systolic.df$ID,
  main="Systolic.df$ID Visual Data Check")

par(ask=TRUE)
plot(Systolic.df$Gender,
  main="Systolic.df$Gender Visual Data Check")

par(ask=TRUE)
plot(Systolic.df$Race,
  main="Systolic.df$Race Visual Data Check",
  col=gray(2:0/2))
  # Note the gray shading scheme and how the values range
  # from 0 to 2 and not 1 to 3.  Counts that begin with 0
  # represent a common means of counting in computer science.

par(ask=TRUE)
plot(density(Systolic.df$Systolic,
  na.rm=TRUE),    # Required for the density() function
  main="Density Plot of Systolic",
  lwd=6, col="red", font.axis=2, font.lab=2)

par(ask=TRUE)
plot(density(Systolic.df$Diastolic,
  na.rm=TRUE),    # Required for the density() function
  main="Density Plot of Diastolic",
  lwd=6, col="red", font.axis=2, font.lab=2)

par(ask=TRUE)
plot(density(Systolic.df$Weight,
  na.rm=TRUE),    # Required for the density() function
  main="Density Plot of Weight",
  lwd=6, col="red", font.axis=2, font.lab=2)
```

The purpose of these initial plots is to gain a general sense of the data and to equally look for outliers. In an attempt to look for outliers, the ylim argument has been avoided, so that all data are plotted. Extreme values may or may not be outliers, but they are certainly interesting and demand attention.

Given that the purpose of this Twoway ANOVA is to examine the variables Gender, Race, and Systolic, graphics will not be included for the remaining variables Diastolic and Weight. Of course, the R-based syntax is this lesson could easily be used as a template for these other variables.

The bar plot is a very common tool for the presentation of graphics, specifically for factor-type variables. Note how the table() function is used to improve on R syntax in tandem with the barplot() function.

```
par(ask=TRUE)
barplot(table(Systolic.df$Gender),
  main="Barplot of Gender",
  xlab="Gender", ylab="Number of Subjects",
  col="red", cex.axis=1.25, cex.names=1.25, cex.lab=1.25)

par(ask=TRUE)
barplot(table(Systolic.df$Race),
  main="Barplot of Race",
  xlab="Race", ylab="Number of Subjects",
  col="red", cex.axis=1.25, cex.names=1.25, cex.lab=1.25)
```

Create an object that represents the output of the table() function against Race and Gender and then use this object instead of keying the table function each time.

```
GenderAndRace <- table(Systolic.df$Race, Systolic.df$Gender)

par(ask=TRUE)
barplot(GenderAndRace,
  beside=FALSE,                                  # Stacked
  main="Stacked Barplot of Gender and Race",
  xlab="Gender by Race", ylab="Number of Subjects",
  col=c("red", "green", "blue"),
  cex.axis=1.25, cex.names=1.25, cex.lab=1.25)
legend("topleft",
  c("Black","Other", "White"),
  col=c("red", "green", "blue"), pch=15, lwd=1)
  # Note the placement of the legend.

par(ask=TRUE)
barplot(GenderAndRace,
  beside=TRUE,                                   # Grouped
  main="Grouped (Side-by-Side) Barplot of Gender and Race",
  xlab="Gender by Race", ylab="Number of Subjects",
  col=c("red", "green", "blue"),
  cex.axis=1.25, cex.names=1.25, cex.lab=1.25)
legend("topleft",  # Note the placement of the legend.
  c("Black","Other", "White"),
  col=c("red", "green", "blue"), pch=15, lwd=1)
  # Note the placement of the legend.
```

The boxplot (e.g., box-and-whiskers plot) is equally a common tool for the presentation of graphics, but for numeric-type variables instead of factor-type variables. As an interesting addition to the boxplot() function, add output from the fivenum() function (Tukey's 5 : minimum, lower-hinge, median, upper-hinge, maximum) in a legend. The, add output from the boxplot.stats() function (lower-whisker, lower-hinge, median, upper-hinge, upper-whisker) as an additional legend. This additional detail will produce a free-standing graphics that is exceptionally information-rich.

6.4 Conduct a Visual Data Check

```
fivenum(Systolic.df$Systolic, na.rm=TRUE)

boxplot.stats(Systolic.df$Systolic)

par(ask=TRUE)
boxplot(Systolic.df$Systolic,
  main="Boxplot of Systolic",
  col="red",   lwd=2, cex.axis=1.25,
  ylab="Systolic", cex.lab=1.25)
savefamily <- par(family="mono") # Courier font
savefont   <- par(font=2)         # Bold
legend("topleft",
  legend = c(
  "> fivenum(Systolic.df$Systolic, na.rm=TRUE)",
  "[1]   78 108 122 140 198                   ",
  "=======================================",
  "Minimum ........................... 078",
  "Lower-Hinge ....................... 108",
  "Median ............................ 122",
  "Upper-Hinge ....................... 140",
  "Maximum ........................... 198"),
  ncol=1, locator(1), xjust=1,
  text.col="darkblue",
  cex=1.05, inset=0.02, bty="n")
par(savefamily)
par(savefont)
savefamily <- par(family="mono") # Courier font
savefont   <- par(font=2)         # Bold
legend("topright",
  legend = c(
  "> boxplot.stats(Systolic.df$Systolic)      ",
  "$stats                                     ",
  "[1]   78 108 122 140 188                   ",
  "=======================================",
  "Lower-Whisker ..................... 078",
  "Lower-Hinge ....................... 108",
  "Median ............................ 122",
  "Upper-Hinge ....................... 140",
  "Upper-Whisker ..................... 188"),
  ncol=1,  locator(1), xjust=1,
  text.col="darkblue",
  cex=1.05,   inset=0.02, bty="n")
par(savefamily); par(savefont)
mtext("The small bubbles indicate outliers.",
  side=1, cex=0.75, font=2)
```

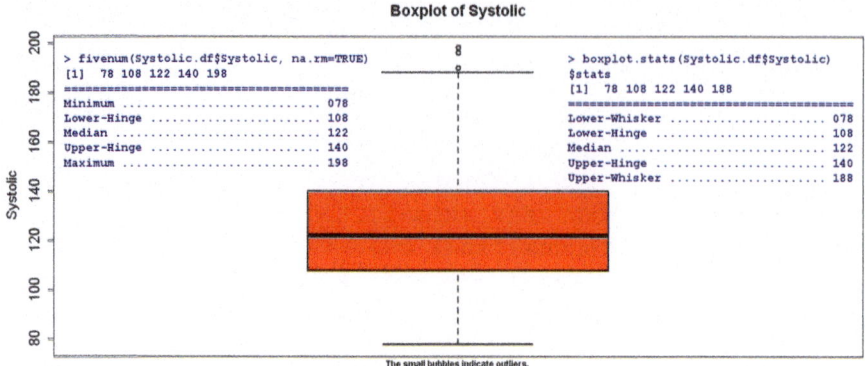

```
par(ask=TRUE)
boxplot(Systolic.df$Systolic ~ Systolic.df$Gender,
  main="Boxplot of Systolic by Gender",
  col=c("pink", "blue"), cex.axis=1.25,
  ylab="Systolic", cex.lab=1.25)
mtext("The small bubbles indicate outliers.",
  side=1, cex=0.75, font=2)

par(ask=TRUE)
boxplot(Systolic.df$Systolic ~ Systolic.df$Race,
  main="Boxplot of Systolic by Race",
  col=c("red", "green", "blue"), cex.axis=1.25,
  ylab="Systolic", cex.lab=1.25)
mtext("The small bubbles indicate outliers.",
  side=1, cex=0.75, font=2)
```

The histogram is another tool for displaying data distribution, typically into organized ranges or *bins*. Note how the lattice package supports fairly detailed histograms, where a great deal of information is conveniently organized into one attractive image.

```
install.packages("lattice")
library(lattice)                # Load the lattice package.
help(package=lattice)           # Show the information page.
sessionInfo()                   # Confirm all attached packages.

par(ask=TRUE)
hist(Systolic.df$Systolic,
  main="Histogram of Systolic",
  xlab="Systolic",  col="red", cex.axis=1.25,
  cex.lab=1.25, font=2)
```

6.4 Conduct a Visual Data Check

```
par(ask=TRUE)
lattice::histogram(~ Systolic | Gender,
  data=Systolic.df, type="percent",
  main="Histogram (lattice::histogram() Function) of Systolic
  by Gender:  Percent",
  xlab=list("Gender", cex=1.15, font=2),
  xlim=c(0,250), ylab=list("Systolic", cex=1.15, font=2),
  layout=c(2,1), col="red")

par(ask=TRUE)
 lattice::histogram(~ Systolic | Gender,
  data=Systolic.df, type="density",
  main="Histogram (lattice::histogram() Function) of Systolic
  by Gender:  Density",
  xlab=list("Gender", cex=1.15, font=2),
  xlim=c(0,250), ylab=list("Systolic", cex=1.15, font=2),
  layout=c(2,1), col="red")

par(ask=TRUE)
lattice::histogram(~ Systolic | Race,
  data=Systolic.df, type="percent",
  main="Histogram (lattice::histogram() Function) of Systolic
  by Race:  Percent",
  xlab=list("Race", cex=1.15, font=2),
  xlim=c(0,250), ylab=list("Systolic", cex=1.15, font=2),
  layout=c(3,1), col="red")

par(ask=TRUE)
lattice::histogram(~ Systolic | Race,
  data=Systolic.df, type="density",
  main="Histogram (lattice::histogram() Function) of Systolic
  by Race:  Density",
  xlab=list("Race", cex=1.15, font=2),
  xlim=c(0,250), ylab=list("Systolic", cex=1.15, font=2),
  layout=c(3,1), col="red")
```

Specialized R-based tools exist that combine multiple types of graphical tools that together generate one convenient and attractive image. Immediately below, the descr::histkdnc() function places into one image a density curve, a normal curve, and a histogram.

```
install.packages("descr")
library(descr)              # Load the descr package.
help(package=descr)         # Show the information page.
sessionInfo()               # Confirm all attached packages.

savelwd       <- par(lwd=4)            # Heavy line
savefont      <- par(font=2)           # Bold
savecex.lab   <- par(cex.lab=1.25)     # Label
savecex.axis  <- par(cex.axis=1.25)    # Axis
par(ask=TRUE)
```

```
descr::histkdnc(Systolic.df$Systolic,
  main="Histogram (descr::histkdnc() Function) of Systolic:
  Superimposed Normal Curve (Blue) and
  Density Curve (Red)",
  xlab="Systolic", col=grey(0.95))  # Allow contrast with lines
par(savelwd); par(savefont); par(savecex.lab);
par(savecex.axis)    # Use ; to move to next line and save space
```

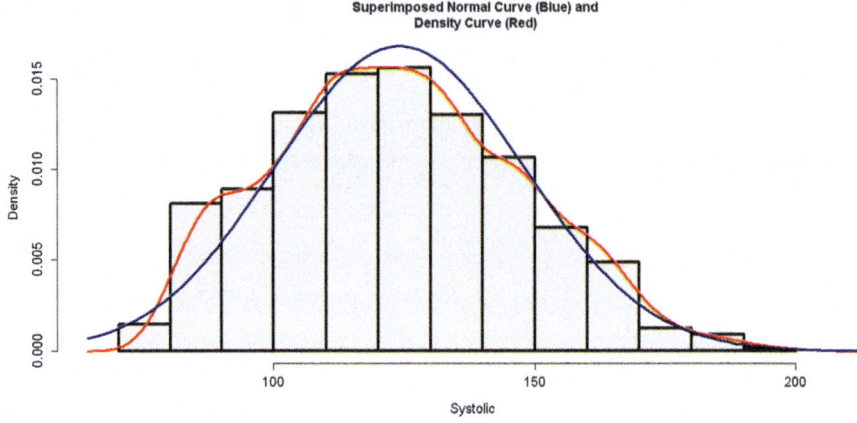

Although the violin plot is not as common as the boxplot, it should be considered as a possible selection for how numerical data are distributed. Note below how the vioplot package and the UsingR package both support functions that generate violin plots.

```
install.packages("vioplot")
library(vioplot)              # Load the vioplot package.
help(package=vioplot)         # Show the information page.
sessionInfo()                 # Confirm all attached packages.

savelwd      <- par(lwd=3)         # Heavy line
savefont     <- par(font=2)        # Bold
savecex.lab  <- par(cex.lab=1.25)  # Label
savecex.axis <- par(cex.axis=1.25) # Axis
par(ask=TRUE)
vioplot::vioplot(Systolic.df$Systolic,
  names=c("Systolic"),  col="red")
title("Violin Plot (vioplot::vioplot() Function) of Systolic")
par(savelwd); par(savefont); par(savecex.lab);
par(savecex.axis)

install.packages("UsingR")
library(UsingR)               # Load the UsingR package.
help(package=UsingR)          # Show the information page.
sessionInfo()                 # Confirm all attached packages.
```

6.5 Descriptive Analysis of the Data

```
savelwd      <- par(lwd=3)           # Heavy line
savefont     <- par(font=2)          # Bold
savecex.lab  <- par(cex.lab=1.25)    # Label
savecex.axis <- par(cex.axis=1.25)   # Axis
par(ask=TRUE)
UsingR::simple.violinplot(Systolic.df$Systolic ~
  Systolic.df$Gender, lty=1, col="red",  ylim=c(0,250))
  title("Violin Plot (UsingR::simple.violinplot() Function)
  of Systolic by Gender")
par(savelwd); par(savefont); par(savecex.lab);
par(savecex.axis)

savelwd      <- par(lwd=3)           # Heavy line
savefont     <- par(font=2)          # Bold
savecex.lab  <- par(cex.lab=1.25)    # Label
savecex.axis <- par(cex.axis=1.25)   # Axis
UsingR::simple.violinplot(Systolic.df$Systolic ~
  Systolic.df$Race, lty=1, col="red", ylim=c(0,250))
  title("Violin Plot (UsingR::simple.violinplot() Function)
  of Systolic by Race")
par(savelwd); par(savefont); par(savecex.lab);
par(savecex.axis)
```

There is seemingly no limit to the number of R-based functions that support graphics and visualization of the data. Experiment as time permits with others.

6.5 Descriptive Analysis of the Data

Given the different ways missing data can impact analyses, it is often helpful to first check for missing data by using the is.na() function and the complete.cases() function against the entire dataset. Both functions return a TRUE or FALSE response, depending on the function and the outcome of whether data are missing or data are not missing.

```
is.na(Systolic.df)             # Check for missing data
complete.cases(Systolic.df)    # Check for complete cases
```

Even for this fairly large dataset, the summary() function may be all that is necessary to gain a sense of the data. As typically used, the summary() function is applied against the entire dataset, thus yielding information about all object variables, including the object variable ID.

```
summary(Systolic.df)
```

Give attention to the listing of NAs, if any, for those object variables with missing data. Again, the summary() function is very useful and it should always be a first selection when preparing descriptive analyses.

Although the summary() function is quite sufficient, descriptive statistics for individual object variables may be desired. To achieve this aim, review the prior lesson *Descriptive Statistics and Measures of Central Tendency* for a comprehensive review of the functions used for descriptive statistics, especially: length(), asbio::Mode(), median(), mean(), sd(), table(), and finally summary(). As needed (but not always, depending on specific functions), the na.rm=TRUE argument or some other similar convention will be used to accommodate missing data.

```
length(Systolic.df)              # N of the dataset
length(Systolic.df$ID)           # N of ID
length(Systolic.df$Gender)       # N of Gender
length(Systolic.df$Race)         # N of Race
length(Systolic.df$Systolic)     # N of Systolic
length(Systolic.df$Diastolic)    # N of diastolic
length(Systolic.df$Weight)       # N of Weight
```

Again, the dataset seems to be in correct form and to conserve space, descriptive statistics, frequency distributions, and measures of central tendency will only be provided for Gender, Race, and Systolic – the focus of this Twoway ANOVA.

```
install.packages("asbio")
library(asbio)              # Load the asbio package.
help(package=asbio)         # Show the information page.
sessionInfo()               # Confirm all attached packages.

asbio::Mode(Systolic.df$Systolic)           # Mode

mode.of.Systolic <-
  names(sort(-table(Systolic.df$Systolic)))[1]
mode.of.Systolic    # Hand calculate mode with NAs
```

Both methods were used to display mode, to show how the asbio::Mode() function provides both modal values for when the data are bimodal. In contrast, the hand calculation displays only the first (e.g., lowest) value representing mode.

```
median(Systolic.df$Systolic, na.rm=TRUE)      # Median

mean(Systolic.df$Systolic, na.rm=TRUE)        # Mean
sd(Systolic.df$Systolic,na.rm=TRUE )          # SD

summary(Systolic.df)
```

Descriptive statistics at the summary level are always useful, but breakout statistics are also needed to gain a more complete understanding of the data. Many functions are presented in this lesson to demonstrate how breakout statistics are obtained when using R. The key here is to discern differences in Systolic Blood Pressure values for the two Gender breakout groups (e.g., Female and Male) and the three Race breakout Groups (e.g., Black, Other, and White).

6.5 Descriptive Analysis of the Data

```
tapply(Systolic, Gender, summary, na.rm=TRUE,
  data=Systolic.df) # Systolic by Gender

par(ask=TRUE) # Use the epicalc package for breakout analyses
epicalc::summ(Systolic.df$Systolic, by=Systolic.df$Gender,
  graph=TRUE, pch=18, ylab="auto",
  main="Sorted Dotplot of Systolic by Gender",
  cex.X.axis=1.15, cex.Y.axis=1.15, font.lab=2, dot.col="auto")
# Note the descriptive statistics and not only the graphic
# that go along with the  epicalc::summ() function.

tapply(Systolic, Race, summary, na.rm=TRUE,
  data=Systolic.df) # Systolic by Gender

par(ask=TRUE) # Use the epicalc package for breakout analyses
epicalc::summ(Systolic.df$Systolic, by=Systolic.df$Race,
  graph=TRUE, pch=18, ylab="auto",
  main="Sorted Dotplot of Systolic by Race",
  cex.X.axis=1.15, cex.Y.axis=1.15, font.lab=2, dot.col="auto")
# Note the descriptive statistics and not only the graphic
# that go along with the  epicalc::summ() function.
```

Again, simple frequency distributions and descriptive statistics are needed to gain a general idea of the data. More detail and greater precision can come later. These initial attempts are simple and are only prepared to provide an initial view of the data.

```
table(Systolic.df$Gender)
table(Systolic.df$Race)
table(Systolic.df$Gender, Systolic.df$Race)
```

The epicalc package is quite useful for those who work in biostatistics and use R. A few of the many functions associated with the epicalc package are shown below, largely to show frequency distributions and percentage representation.

```
epicalc::tableStack(Gender, dataFrame=Systolic.df,
  by="none", count=TRUE, decimal=2,
  percent=c("column", "row"), frequency=TRUE,
  name.test=FALSE, total.column=TRUE, test=FALSE)

epicalc::tableStack(Race, dataFrame=Systolic.df,
  by=Gender, count=TRUE, decimal=2,
  percent=c("column", "row"), frequency=TRUE,
  name.test=FALSE, total.column=TRUE, test=FALSE)

epicalc::tableStack(Race, dataFrame=Systolic.df,
  by="none", count=TRUE, decimal=2,
  percent=c("column", "row"), frequency=TRUE,
  name.test=FALSE, total.column=TRUE, test=FALSE)
```

```
epicalc::tableStack(Gender, dataFrame=Systolic.df,
  by=Race, count=TRUE, decimal=2,
  percent=c("column", "row"), frequency=TRUE,
  name.test=FALSE, total.column=TRUE, test=FALSE)

par(ask=TRUE)
epicalc::tab1(Systolic.df$Gender,   # Bar Plot
  decimal=2,                        # Use the tab1() function
  sort.group=FALSE,                 # from the epicalc
  cum.percent=TRUE,                 # package to see details
  graph=TRUE,                       # about the selected
  missing=TRUE,                     # object variable. (The
  bar.values=c("frequency"),        # 1 of tab1 is the one
  horiz=FALSE,                      # numeric character and
  cex=1.15,                         # it is not the letter
  cex.names=1.15,                   # l).
  cex.lab=1.15, cex.axis=1.15,
  main="Factor Levels for Object Variable Gender",
  ylab="Frequency of Gender, Includings NAs if Any",
  col= c("pink", "blue"), gen=TRUE)

par(ask=TRUE)
epicalc::tab1(Systolic.df$Race,     # Bar Plot
  decimal=2,                        # Use the tab1() function
  sort.group=FALSE,                 # from the epicalc
  cum.percent=TRUE,                 # package to see details
  graph=TRUE,                       # about the selected
  missing=TRUE,                     # object variable. (The
  bar.values=c("frequency"),        # 1 of tab1 is the one
  horiz=FALSE,                      # numeric character and
  cex=1.15,                         # it is not the letter
  cex.names=1.15,                   # l).
  cex.lab=1.15, cex.axis=1.15,
  main="Factor Levels for Object Variable Race",
  ylab="Frequency of Race, Includings NAs if Any",
  col= c("red", "green", "blue"), gen=TRUE)
```

The catspec package and specifically the catspec::ctab() function support a very rich display of frequency distributions. This resource should be used, especially if there is a desire to copy and past the frequency distribution table from R into a word processed technical report.

```
install.packages("catspec")
library(catspec)              # Load the catspec package.
help(package=catspec)         # Show the information page.
sessionInfo()                 # Confirm all attached packages.

catspec::ctab(Systolic.df$Gender,
  dec.places=2,
  type=c("n", "row", "total"),
```

6.5 Descriptive Analysis of the Data

```
    style="wide", percentages=TRUE,
    addmargins=TRUE)

catspec::ctab(Systolic.df$Race,
    dec.places=2,
    type=c("n", "row", "total"),
    style="wide", percentages=TRUE,
    addmargins=TRUE)

catspec::ctab(Systolic.df$Gender, Systolic.df$Race,
    dec.places=2,
    type=c("n", "row", "column", "total"), style="long",
    percentages=TRUE, addmargins=TRUE)
```

```
                Var2    Black    Other    White      Sum
Var1
Female Count            244.00   163.00   393.00   800.00
       Row %             30.50    20.38    49.12   100.00
       Column %          40.67    40.75    39.30   120.72
       Total %           12.20     8.15    19.65    40.00
Male   Count            356.00   237.00   607.00  1200.00
       Row %             29.67    19.75    50.58   100.00
       Column %          59.33    59.25    60.70   179.28
       Total %           17.80    11.85    30.35    60.00
Sum    Count            600.00   400.00  1000.00  2000.00
       Row %             60.17    40.12    99.71   200.00
       Column %         100.00   100.00   100.00   300.00
       Total %           30.00    20.00    50.00   100.00
```

```
catspec::ctab(Systolic.df$Race, Systolic.df$Gender,
    dec.places=2,
    type=c("n", "row", "column", "total"), style="long",
    percentages=TRUE, addmargins=TRUE)
```

The vcd::structable() function is equally useful for the display of increasingly complex frequency distributions.

```
install.packages("vcd")
library(vcd)              # Load the vcd package.
help(package=vcd)         # Show the information page.
sessionInfo()             # Confirm all attached packages.

vcd::structable(Gender ~ Race, data=Systolic.df)

vcd::structable(Race ~ Gender, data=Systolic.df)
```

As is nearly always the case with R, there is no one-and-only-one way to obtain breakout statistics. Look at the Boolean approach shown below and the way use of == (== and not = and equally there is no space between the two == characters) for selection purposes has been used, to obtain Systolic breakout statistics first by

Gender, then by Race, and then by Gender and Race. There are no missing values for this data set so it is not necessary to use na.rm=TRUE or any other accommodation for missing values (e.g., NA).

```
mean(Systolic.df$Systolic)

mean(Systolic.df$Systolic[Systolic.df$Gender=="Female"])
mean(Systolic.df$Systolic[Systolic.df$Gender=="Male"])
mean(Systolic.df$Systolic[Systolic.df$Race=="Black"])
mean(Systolic.df$Systolic[Systolic.df$Race=="Other"])
mean(Systolic.df$Systolic[Systolic.df$Race=="White"])

mean(Systolic.df$Systolic[Systolic.df$Gender=="Female" &
   Systolic.df$Race=="Black"])
mean(Systolic.df$Systolic[Systolic.df$Gender=="Female" &
   Systolic.df$Race=="Other"])
mean(Systolic.df$Systolic[Systolic.df$Gender=="Female" &
   Systolic.df$Race=="White"])

mean(Systolic.df$Systolic[Systolic.df$Gender=="Male" &
   Systolic.df$Race=="Black"])
mean(Systolic.df$Systolic[Systolic.df$Gender=="Male" &
   Systolic.df$Race=="Other"])
mean(Systolic.df$Systolic[Systolic.df$Gender=="Male" &
   Systolic.df$Race=="White"])
```

The number of R-based functions used for descriptive statistics is certainly large, but that should be expected given the importance of descriptive statistics. Below, consider the use of the fields package and the tables package.

```
install.packages("fields")
library(fields)              # Load the fields package.
help(package=fields)         # Show the information page.
sessionInfo()                # Confirm all attached packages.

fields::stats(Systolic.df$Systolic)

fields::stats(Systolic.df$Systolic, by=Systolic.df$Gender)
fields::stats(Systolic.df$Systolic, by=Systolic.df$Race)

install.packages("tables")
library(tables)              # Load the tables package.
help(package=tables)         # Show the information page.
sessionInfo()                # Confirm all attached packages.

tables::tabular( (Gender + 1) ~ (n=1) + Format(digits=2)*
   (Systolic + Diastolic + Weight)*(min + max + mean + sd),
   data=Systolic.df )
```

```
tables::tabular( (Race + 1) ~ (n=1) + Format(digits=2)*
  (Systolic + Diastolic + Weight)*(min + max + mean + sd),
  data=Systolic.df )
```

Many other packages and functions from these packages could be presented, but are excluded from this lesson to conserve space. If time permits, look into the etable::tabular.ade() function and the descr::compmeans() function. Other functions that serve this purpose exist and formal searches at RSeek (http://rseek.org) should be useful for this purpose.

6.6 Conduct the Statistical Analysis

The preceding graphical images and descriptive statistics, both summary descriptive statistics and breakout descriptive statistics, provide a fairly good idea of Systolic Blood Pressure statistics, overall and by Gender and Race:

```
Systolic Blood Pressure by Gender and by Race
```

Gender	Race	N	MIN	MAX	MEAN	SD
Female	Black	244	78.0	198	129	24.4
Female	Other	163	80.0	188	123	22.8
Female	White	393	80.0	190	122	24.1
Female	Total	800	78.0	198	125	24.1
Male	Black	356	80.0	198	130	22.7
Male	Other	237	80.0	182	120	24.0
Male	White	607	78.0	190	121	23.1
Male	Total	1200	78.0	198	124	23.5

The data have now been brought into this R session and reviewed using graphical images, frequency distributions, and measures of central tendency, with presentation at the summary level and breakout levels. Given that these many actions are acceptable, the data are ready for inferential analyses, such as Twoway ANOVA in this lesson.

The task now is to use Twoway ANOVA to determine if there are statistically significant differences (p <= 0.05) in Systolic Blood Pressure between (1) the two Gender breakout groups (e.g., Female and Male), (2) the three breakout groups for Race (e.g., Black, Other, and White), and any possible interaction between Gender and Race. Remember that as useful as the graphical images and descriptive statistics may be, an inferential test is needed for final determination of statistically significant differences.

R supports many possible ways to perform a Two-Way ANOVA. A few different methods for Twoway ANOVA are shown below.

Use the formula for a Two-Way Factorial Design ANOVA, which is typically based on the aov() function and is represented as:

```
# fit1 <- aov(y ~ A + B + A:B, data=dataframe)
# summary(fit1)

# fit2 <- aov(y ~ A*B, data=dataframe)
# summary(fit2)
```

```
y   = Measured datum, (e.g., Weight, Exam Score, etc.)
A   = Factor Variable A (e.g., Gender, Race-Ethnicity, etc.)
B   = Factor Variable B (e.g., Soil Type, Breed Type, etc.)
A:B = Interaction of Factor Variable A and Factor Variable B
```

Both ANOVA formulas yield the same result.

Twoway ANOVA Method 1

```
fit1 <- aov(y ~ A + B + A:B, data=dataframe)

Systolic.GenderRace.fit1 <-
  aov(Systolic ~ Gender + Race + Gender:Race,
  data=Systolic.df)

summary(Systolic.GenderRace.fit1)
```

```
> summary(Systolic.GenderRace.fit1)
              Df   Sum Sq Mean Sq F value   Pr(>F)
Gender         1      458     458   0.831    0.362
Race           2    28221   14110  25.617 1.04e-11 ***
Gender:Race    2      735     368   0.668    0.513
Residuals   1994  1098329     551
---
Signif. codes:  0 *** 0.001 ** 0.01 * 0.05 0.1 1
>
```

Twoway ANOVA Method 2

```
fit2 <- aov(y ~ A*B, data=dataframe)

Systolic.GenderRace.fit2 <-
  aov(Systolic ~ Gender*Race,
  data=Systolic.df)

summary(Systolic.GenderRace.fit2)
```

```
> summary(Systolic.GenderRace.fit2)
              Df Sum Sq Mean Sq F value   Pr(>F)
Gender         1    458     458   0.831    0.362
Race           2  28221   14110  25.617 1.04e-11 ***
Gender:Race    2    735     368   0.668    0.513
```

6.6 Conduct the Statistical Analysis

```
Residuals    1994 1098329      551
---
Signif. codes:  0 *** 0.001 ** 0.01 * 0.05  0.1  1
>
```

The immediate interpretation of this Twoway ANOVA for Systolic Blood Pressure by Gender, by Race, and interaction by Gender and Race is that:

- There is no statistically significant difference in Systolic Blood Pressure by Gender (p <= 0.05). The calculated p value (p <= 0.362) exceeds the criterion p value (p <= 0.05) and therefore the Null Hypothesis is confirmed for this part of the Twoway ANOVA.
- However, there is a statistically significant difference in Systolic Blood Pressure by Race (p <= 0.05). The calculated p value (p <= 1.04e-11, note the use of e-notation) is indeed less than the criterion p value (p <- 0.05) and therefore the Null Hypothesis is rejected (e.g., not accepted) for this part of the Twoway ANOVA. For further confirmation of this finding of significance, look at the three * characters immediately after the p value for Race and compare these three * characters to the significance codes listing immediately below the ANOVA table.
- There is no statistically significant interaction for Systolic Blood Pressure by Gender and by Race (p <= 0.05). The calculated p value (p <= 0.513) exceeds the criterion p value (p <= 0.05) and therefore a significant interaction cannot be confirmed.

To gain a sense of the descriptive statistics (summary and breakout), use the model.tables() function for another view of Grand Mean and Mean and N ("rep" for replication in this output) for each cell in the factorial table.

```
print(model.tables(Systolic.GenderRace.fit1,"means"), digits=5)
print(model.tables(Systolic.GenderRace.fit2,"means"), digits=5)

> print(model.tables(Systolic.GenderRace.fit2,"means"), digits=5)
Tables of means
Grand mean

123.934

 Gender
    Female     Male
    124.52   123.54
rep 800.00  1200.00

 Race
     Black   Other    White
    129.67  121.18   121.59
rep 600.00  400.00  1000.00

 Gender:Race
         Race
```

```
Gender    Black  Other  White
  Female 129.15 122.60 122.44
  rep    244.00 163.00 393.00
  Male   130.03 120.22 121.03
  rep    356.00 237.00 607.00
>
```

As shown in the output from application of the model.tables() function, the Grand mean (e.g., the mean for all subjects) for Systolic Blood Pressure is 123.94. The breakout means, however, are perhaps of most interest in this lesson.

- There was no statistically significant difference (p <= 0.05) in Systolic Blood Pressure by Gender and perhaps not surprisingly, the two means for Systolic Blood Pressure by Gender are: Female = 124.52 and Male = 123.54.
- There was a statistically significant difference (p <= 0.05) in Systolic Blood Pressure by Race and correspondingly, the three means for Systolic Blood Pressure by Race are: Black = 129.67, Other = 121.18, and White = 121.59.

The mean Systolic Blood Pressure for Black subjects greatly exceeds Systolic Blood Pressure mean values for Other subjects and White subjects and it is now known that there is a statistically significant difference (p <= 0.05). More information is needed, however, to make finite declarations of findings.

As a throwaway diagnostic, use the plot.design() function to see general trends for Systolic counts by each breakout group. In this figure, the mean difference in Systolic Blood Pressure for Black subjects, compared to Other subjects and White subjects, is quite obvious but again, an inferential test such as Twoway ANOVA is needed for final determination of difference.

```
par(ask=TRUE)
plot.design(Systolic ~ Gender + Race,
   data=Systolic.df,
   main="Systolic Blood Pressure (Mean) by Gender and by Race",
   lwd=3, font=2, cex.lab=1.25)
```

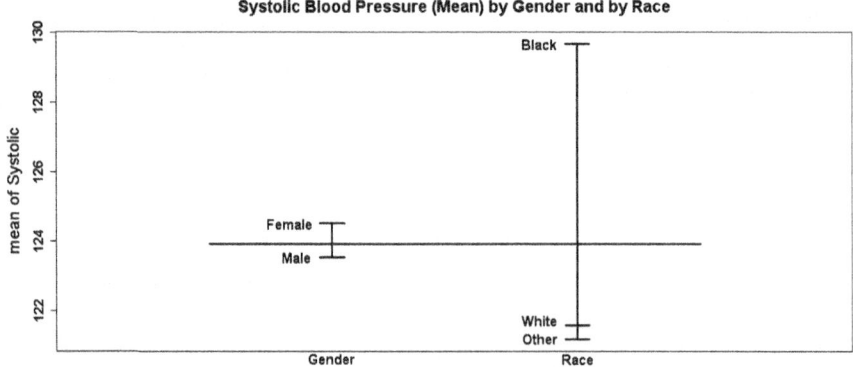

6.6 Conduct the Statistical Analysis

Recall that interactions can possibly mask main effects, but there was no statistically significant (p <= 0.05) interaction in this example. Even so, an interaction plot is a useful tool for visualizing possible interactions and to visualize outcomes from multiple perspectives.

```
savelwd       <- par(lwd=4)           # Heavy line
savefont      <- par(font=2)          # Bold
savecex.lab   <- par(cex.lab=1.25)    # Label
savecex.axis  <- par(cex.axis=1.25)   # Axis
par(ask=TRUE)
interaction.plot(Systolic.df$Gender, Systolic.df$Race,
   Systolic.df$Systolic,   # Note the ordering of variables.
   main="Interaction Plot: Gender, Race, and Systolic",
   fun=mean,               # Use mean instead of median.
   legend=TRUE, trace.label="Race", fixed=TRUE,
   col=c("red", "green", "blue"),
   lwd=4, lty=c("solid", "dashed", "dotdash"),
   xlab=" ",               # Blank label to allow for output.
   ylab="Systolic", font.lab=2, ylim=c(115,135), xtick=TRUE)
par(savelwd)       # Return to original setting.
par(savefont)      # Return to original setting.
par(savecex.lab)   # Return to original setting.
par(savecex.axis)  # Return to original setting.

savelwd       <- par(lwd=4)           # Heavy line
savefont      <- par(font=2)          # Bold
savecex.lab   <- par(cex.lab=1.25)    # Label
savecex.axis  <- par(cex.axis=1.25)   # Axis
par(ask=TRUE)
interaction.plot(Systolic.df$Race, Systolic.df$Gender,
   Systolic.df$Systolic,   # Note the ordering of variables.
   main="Interaction Plot: Race, Gender, and Systolic",
   fun=mean, # Use mean instead of median.
   legend=TRUE, trace.label="Gender", fixed=TRUE,
   col= c("pink", "blue"),
   lwd=4, lty=c("solid", "dashed"),
   xlab=" ",               # Blank label to allow for output.
   ylab="Systolic", font.lab=2, ylim=c(115,135), xtick=TRUE)
par(savelwd)       # Return to original setting.
par(savefont)      # Return to original setting.
par(savecex.lab)   # Return to original setting.
par(savecex.axis)  # Return to original setting.
```

6.7 Summary

The graphics and statistics in this lesson provide a great deal of information, for this one sample, about Systolic Blood Pressure with emphasis on Gender and Race. For a complete summary, it is perhaps best to first revisit the Null Hypothesis and the breakout analyses deriving from use of the aov() function and the model.tables() function:

Null Hypothesis (Ho): There is no statistically significant difference (p <= 0.05) in Systolic Blood Pressure by Gender (Female and Male), by Race (Black, Other, White), and by the interaction of Gender and Race.

Focusing on the Null Hypothesis and the aov() function, the Twoway ANOVA table confirmed that there was a statistically significant difference (p <= 0.05) in Systolic Blood Pressure by Race but there was no statistically significant difference (p <= 0.05) in Systolic Blood Pressure by either Gender or by the interaction of Gender and Race. Again, revisit the prior finding that:

- There was no statistically significant difference (p <= 0.05) in Systolic Blood Pressure by Gender (Female = 124.52 and Male = 123.54).
- There was a statistically significant difference (p <= 0.05) in Systolic Blood Pressure by Race (Black = 129.67, Other = 121.18, and White = 121.59.
- There is no statistically significant interaction (p <= 0.05) of Systolic Blood Pressure by Gender and by Race.

To gain more precision, however (especially to determine more about differences by Race, where significant difference was found), apply the post hoc TukeyHSD() function, which is of course associated with Oneway ANOVA, against Systolic and Race. Then, use the plot() function against the TukeyHSD() function to visually reinforce findings.

```
Systolic.by.Race.OnewayANOVA <- aov(Systolic ~ Race,
   data=Systolic.df)
   # Oneway ANOVA of Systolic by Race

Systolic.by.Race.OnewayANOVA

summary(Systolic.by.Race.OnewayANOVA)

TukeyHSD(Systolic.by.Race.OnewayANOVA)         # Apply TukeyHSD()

> TukeyHSD(Systolic.by.Race.OnewayANOVA)   # Apply TukeyHSD()
  Tukey multiple comparisons of means
    95% family-wise confidence level

Fit: aov(formula = Systolic ~ Race, data = Systolic.df)

Race
                  p adj
Other-Black   0.0000001
```

6.7 Summary

```
White-Black   0.0000000
White-Other   0.9556985
>
```

The p adj column of the TukeyHSD() function provides evidence that there is a statistically significant difference (p <= 0.05) for the following Race comparisons: Other-Black and White-Black. Note how the p values for these two combinations are both <= 0.05.

In contrast, the p value for White-Other, shown above as 0.9556985, exceeds p <= 0.05 and accordingly there is no difference in Systolic Blood Pressure between White subjects and Other subjects.

Finally, plot the TukeyHSD findings, to visually reinforce where there are differences in Systolic Blood Pressure by Race and where there are not.

```
savelwd       <- par(lwd=4)           # Heavy line
savefont      <- par(font=2)          # Bold
savecex.lab   <- par(cex.lab=1.25)    # Label
savecex.axis  <- par(cex.axis=1.25)   # Axis
par(ask=TRUE)
plot(TukeyHSD(Systolic.by.Race.OnewayANOVA),
   las=3, col.axis="darkblue", col="red")
par(savelwd)         # Return to original setting.
par(savefont)        # Return to original setting.
par(savecex.lab)     # Return to original setting.
par(savecex.axis)    # Return to original setting.
```

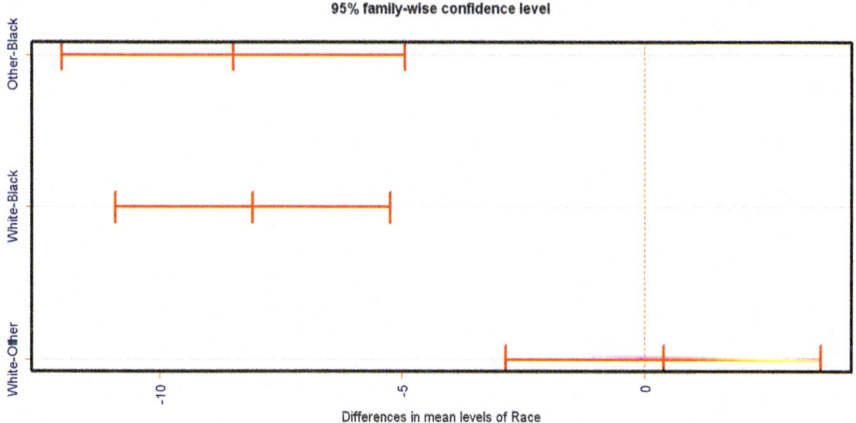

In summary, this Twoway ANOVA provided evidence that Gender had no impact on Systolic Blood Pressure, Race had an impact on Systolic Blood Pressure with Black subjects showing significantly higher Systolic Blood Pressure readings than their counterparts, and there was no observed interaction between Gender and Race. To be more precise:

- There was no statistically significant difference ($p <= 0.05$) in Systolic Blood Pressure between Female subjects and Male subjects.
- There was, however, a statistically significant difference ($p <= 0.05$) in Systolic Blood Pressure by Race.
  - There was an observed difference between Black subjects and Other subjects and there was also an observed difference between Black subjects and White subjects. In both cases, Black subjects had Systolic Blood Pressure values that were significantly higher than the Systolic Blood Pressure values of their counterparts from the two other Race breakout groups.
  - There was no observed difference in Systolic Blood Pressure between White subjects and Other subjects.
- Finally, there was no statistically significant interaction ($p <= 0.05$) in Systolic Blood Pressure by Gender and Race.

When viewing these conclusions, observe the caution that replication and rigid attention to established protocols are inherent to the research process. This sample is merely one attempt in this overall assessment of factors impacting health-related concerns and concomitantly biostatistics.

6.8 Addendum: Other Packages for Display of Twoway ANOVA

R has the advantage that the user community has contributed more than 3,000 packages to supplement the many functions available in the base package, when R is first downloaded. As expected, Twoway ANOVA is the focus of some of these additional packages. A few of these other R functions will be demonstrated below. The information gained from these additional R functions provides a rich understanding of the data and they provide additional insight into interaction(s), when present. Experienced researchers purposely use redundant approaches to test data, to gain perspective from multiple viewpoints.

```
install.packages("s20x")
library(s20x)              # Load the s20x package.
help(package=s20x)         # how the information page.
sessionInfo()              # Confirm all attached packages.

par(ask=TRUE)
s20x::boxqq(Systolic ~ Gender, data=Systolic.df)

par(ask=TRUE)
s20x::boxqq(Systolic ~ Race, data=Systolic.df)
```

6.8 Addendum: Other Packages for Display of Twoway ANOVA

The s20x::interactionPlots() function does not work correctly if there are missing data. Of course, this issue is not a concern in this sample. If it were, the dataset would need to be adjusted so that each case has a full set of data but equally recall that this action is not always desirable.

```
par(ask=TRUE)
s20x::interactionPlots(Systolic ~ Gender+Race,
  Systolic.df,
  xlab="Gender", xlab2="Race",
  ylab="Systolic Blood Pressure",
  type="hsd", tick.length=0.1,
  interval.distance=0.1, col.width=3/4,
  xlab.distance=1, xlen=1.25, ylen=1.25)
# There are four options to the type= argument:
# tukey, hsd, lsd, and ci.
```

Going back to the prior creation of Systolic.GenderRace.fit1 and Systolic.GenderRace.fit2, notice how the s20x::summary2way() function can be used to generate a Two-Way ANOVA table.

```
s20x::summary2way(Systolic.GenderRace.fit1,
  page="table", digit=5, conf.level=0.95,
  print.out=TRUE)

s20x::summary2way(Systolic.GenderRace.fit1,
  page="means", digit=5, conf.level=0.95,
  print.out=TRUE)

s20x::summary2way(Systolic.GenderRace.fit2,
  page="table", digit=5, conf.level=0.95,
  print.out=TRUE)

s20x::summary2way(Systolic.GenderRace.fit2,
  page="means", digit=5, conf.level=0.95,
  print.out=TRUE)
```

```
Cell-means Matrix:
                          Race
                 Black   Other   White   Gender
  Gender Female  129.15  122.60  122.44  124.73
         Male    130.03  120.22  121.03  123.76
         Race    129.59  121.41  121.74  124.25

Numeric Summary:
              Size   Mean       Median  Std Dev   Midspread
All Data      2000   123.934    122     23.75192  32
By  Gender :
Female        800    124.52     122     24.09742  32
Male          1200   123.54333  124     23.52075  32.5
```

```
By Race :
Black              600      129.67333 128      23.42865 34
Other              400      121.19    118      23.5425  34
White              1000     121.588   121      23.45371 34
Combinations:
Female.Black       244      129.14754 128      24.42884 34.5
Female.Other       163      122.60123 120      22.8426  30
Female.White       393      122.44275 122      24.05897 34
Male.Black         356      130.03371 130      22.7456  32.5
Male.Other         237      120.21941 116      24.0117  34
Male.White         607      121.0346  120      23.05648 34
```

Give special attention to the page="means" argument and how it provides detailed summary statistics for: overall, by the different factor object variables, and by combinations of the different factor object variables. This printout is quite valuable and may be of interest beyond statistics in the ANOVA table.

The output from use of the s20x package is especially rich and it should always be considered, both for the Two-Way ANOVA output as well as the way breakout descriptive statistics are presented.

Revisit the previously created objects Systolic.GenderRace.fit1 and Systolic.GenderRace.fit2 and use them again, now with the car package. Although it is beyond the purpose of this lesson, as time permits review the different models commonly used with ANOVA and equally, review the many designs that are possible with ANOVA. Specifically, review documentation for the car::Anova() function to see possible cautions with type="III" and why it should be used only after careful consideration.

```
install.packages("car")
library(car)                 # Load the car package.
help(package=car)            # Show the information page.
sessionInfo()                # Confirm all attached packages.

car::Anova(Systolic.GenderRace.fit1)
car::Anova(Systolic.GenderRace.fit1, type="II")
car::Anova(Systolic.GenderRace.fit1, type="III")

car::Anova(Systolic.GenderRace.fit2)
car::Anova(Systolic.GenderRace.fit2, type="II")
car::Anova(Systolic.GenderRace.fit2, type="III")
```

6.9 Prepare to Exit, Save, and Later Retrieve This R Session

It is common to prepare R syntax in a separate file, using a simple ASCII text editor. If time permits, experiment with Crimson Editor, Tinn-R, or vim, but there are many other possible selections.

6.9 Prepare to Exit, Save, and Later Retrieve This R Session

```
getwd()              # Identify the current working directory.
ls()                 # List all objects in the working
                     # directory.
ls.str()             # List all objects, with finite detail.
list.files()         # List files at the PC directory.

save.image("R_Lesson_Twoway_ANOVA.rdata")

getwd()              # Identify the current working directory.
ls()                 # List all objects in the working
                     # directory.
ls.str()             # List all objects, with finite detail.
list.files()         # List files at the PC directory.

alarm()              # Alarm, notice of upcoming action.
q()                  # Quit this session.
                     # Prepare for Save workspace image? query.
```

Use the R Graphical User Interface (GUI) to load the saved rdata file: **File** and then **Load Workspace**. Otherwise, use the load() function, keying the full pathname, to load the .rdata file and retrieve the session.

Recall, however, that it may be just as useful to simply use a R script file (typically saved as a .txt ASCII-type file) and recreate the analyses and graphics, provided the data files remain available.

Chapter 7
Correlation and Linear Regression

Abstract The purpose of this lesson on correlation and linear regression is to provide guidance on how R can be used to determine the association between two variables and to then use this degree of association to predict future outcomes. Past behavior is the best predictor of future behavior. This concept applies in the biological sciences, physical sciences, social sciences, and also in economics. By knowing past relationships between variables (e.g., correlation), it is then possible to build a prediction equation to foretell future values for selected variables (e.g., regression). This lesson will focus on Pearson's Product Moment Coefficient of Correlation (Pearson's r, perhaps the most common test for determining if there is an association between phenomena) and linear regression.

Keywords Association • Coefficient of correlation • Correlation • Linear regression • Minimal Adequate Model (MAM) • Pearson's r • Regression • Regression line • Scatter plot • Scatter plot matrix • SPLOM • Stepwise regression

7.1 Background on This Lesson

7.1.1 Description of the Data

This lesson on correlation (e.g., association) and linear regression focused on livestock in a contemporary feeder operation as the biological specimen in question. This lesson has been prepared to determine if there is a correlation in final weight by measures associated with vigor (e.g., a measure of general health and appearance). The data are from a sample of 3,308 animals and there are some missing data.

Some livestock operations use a *farm to fork* approach, where animals are born and over time raised for slaughter all at one location, under the control of a single management team. In contrast, this sample is based on a specialized feeder approach, where livestock are raised to a certain weight by one operator, sold, and then moved to a specialized location where they are fed, managed, and finally finished-out to a desired weight suitable for slaughter.

This lesson is specific to how R is used to determine measures of correlation and linear regression. If needed, review the many materials easily available on the nature of correlation and how Pearson's r ranges from -1.0 (*perfect* negative or declining association) to $+1.0$ (*perfect* positive or ascending association).

As a brief example of correlation, consider the following scenario that borrows from public health. Assume that at the national level, there is a negative correlation between miles of paved roads per capita and the rate (e.g., percentage) of infant mortality:

- Let X equal miles of paved roads per capita.
- Let Y equal the rate of infant mortality.

In this scenario, as X increases Y decreases. That is to say, as the miles of paved roads per capita in a country increases the rate of infant mortality (e.g., the percentage of babies who die at or soon after birth) decreases.

Why does this occur? Well, that is a totally different concern. Do not assume that cause and effect (e.g., causation) is in place here. Sure, a few mothers may get to the hospital faster, and therefore have children who survive birth, if they can drive on a paved road as opposed to the use of unreliable transportation on unpaved roads. but that is hardly the reason for the broad association between roads and healthy babies. Other factors, such as economic wealth, societal issues, and general development of infrastructure are also prime concerns that impact this public health measure. It could be stated that miles of paved roads per capita only serves as a *proxy* for public health and wellness.

Other than subject identification number, there are no factor-type data in this lesson. Weights are presented as pounds (e.g., Lbs.). Vigor is viewed as a numeric measure of general health and appearance, using a scale that ranges from Low Vigor to High Vigor, or a scale of 0.00 to 10.00.

7.1.2 Null Hypothesis (Ho)

Because there are more than a few correlations associated with this dataset (e.g., X:X, X:Y, X:Z, etc.), a generic Null Hypothesis (Ho) is presented below:

There is no statistically significant correlation (p <= 0.05) between Variable X and Variable Y.

7.2 Data Import of a .csv Spreadsheet-Type Data File into R

For this lesson, the dataset has been prepared in .csv (e.g., comma-separated values) file format. The data are separated by commas. The data are not separated by tabs and the data are not separated by spaces.

```
###############################################################
# Housekeeping                          Use for All Analyses  #
###############################################################
date()              # Current system time and date.
R.version.string    # R version and version release date.
ls()                # List all objects in the working
                    # directory.
rm(list = ls())     # CAUTION: Remove all files in the working
                    # directory. If this action is not desired,
                    # use the rm() function one-by-one to remove
                    # the objects that are not needed.
ls.str()            # List all objects, with finite detail.
getwd()             # Identify the current working directory.
setwd("F:/R_Biostatistics")
                    # Set to a new working directory.
                    # Note the single forward slash and double
                    # quotes.
                    # This new directory should be the directory
                    # where the data file is located, otherwise
                    # the data file will not be found.
getwd()             # Confirm the working directory.
list.files()        # List files at the PC directory.
###############################################################

LStockVg.df <- read.table (file =
  "Livestock_Vigor.csv",
  header = TRUE,
  sep = ",")                     # Import the .csv file

getwd()                          # Identify the working directory
ls()                             # List objects
attach(LStockVg.df)              # Attach the data, for later use
str(LStockVg.df)                 # Identify structure
nrow(LStockVg.df)                # List the number of rows
ncol(LStockVg.df)                # List the number of columns
dim(LStockVg.df)                 # Dimensions of the data frame
names(LStockVg.df)               # Identify names
colnames(LStockVg.df)            # Show column names
rownames(LStockVg.df)            # Show row names
head(LStockVg.df)                # Show the head
tail(LStockVg.df)                # Show the tail
LStockVg.df                      # Show the entire data frame
summary(LStockVg.df)             # Summary statistics
```

An object called LStockVg.df has now been created and accommodated. This R-based object is a data frame and it consists of the data originally included in the file Livestock_Vigor.csv, a comma-separated .csv file. To avoid possible conflicts, make sure that there are no prior R-based objects called LStockVg.df. The prior use of rm(list=ls()) accommodates this concern, removing all prior objects in the current R session.

Note how it was only necessary to key the filename for the .csv file and not the full pathname since the R working directory is currently set to the directory and/or subdirectory where this .csv file is located (see the Housekeeping section at the beginning of this lesson).

7.3 Organize the Data and Display the Code Book

Now that the data have been imported into R, it is usually necessary to check the data for format and then make any changes that may be needed, to organize the data. This dataset is fairly large (N = 3,308 subjects) but the data are fairly simple in terms of organization. Since the focus is on numeric-type object variables, only.

As a new feature that has not yet been introduced in this set of lessons, a rowname will be used with this dataset, using the rownames() function. Although this action is not required, rownames may be helpful when working with increasingly large datasets. The rownames() function assigns a unique identifier for each row in this dataset of 3,308 rows, each beginning with the term Animal in this example.

```
rownames(LStockVg.df) <- paste('Animal', 1:3308)

tail(LStockVg.df)   # Show the tail, now to confirm rownames
```

For this lesson, the class() function, str() function, and duplicated() function will be used to be sure that data are organized as desired.

```
class(LStockVg.df)
class(LStockVg.df$Subject)
class(LStockVg.df$WeightInitial)
class(LStockVg.df$WeightFinish)
class(LStockVg.df$VigorInitial)
class(LStockVg.df$Vigor100Lbs)
class(LStockVg.df$Vigor200Lbs)
class(LStockVg.df$VigorFinish)
```

Following along with all prior lessons and general standards of good programming practices, DataFrame$ObjectName notation has been used for object variables.

```
str(LStockVg.df)            # Structure

duplicated(LStockVg.df)     # Duplicates
```

7.3 Organize the Data and Display the Code Book

The class for each object seems to be correct and there are no duplicate rows of data in the data frame. Saying this, a Code Book will help with future understanding this dataset.

A Code Book is an essential aid for anyone involved in the day-to-day activities of the research and statistics process. The Code Book is typically brief and only serves as a useful reminder for what can be easily forgotten months (or even weeks) later, to make it easy to decipher what may otherwise be seen as arcane numeric codes. Coding schemes that are intuitively obvious today can easily become obtuse tomorrow.

Now that the class(), str(), and duplicated() functions have been used for basic diagnostics, consult the Code Book and coerce each object, as needed, into its correct class.

```
############################################################
# Code Book                                                #
############################################################
#                                                          #
# Subject ................... Factor (e.g., nominal)       #
#              A unique ID ranging from S0001 to S3308     #
#                                                          #
# WeightInitial .............Numeric (e.g., interval)      #
#                 Initial feeder stock weight, with        #
#                        average weight about 35 lbs       #
#                                                          #
# WeightFinish .............Numeric (e.g., interval)       #
#                 Finished feeder stock weight, with       #
#                       average weight about 250 lbs       #
#                                                          #
# VigorInitial ............ Numeric (e.g., interval)       #
#         A measure of general health and appearance,      #
#                        ranging from 0.00 to 10.00        #
#                                                          #
# Vigor100Lbs ............. Numeric (e.g., interval)       #
#         A measure of general health and appearance,      #
#                        ranging from 0.00 to 10.00        #
#                                                          #
# Vigor200Lbs ............. Numeric (e.g., interval)       #
#         A measure of general health and appearance,      #
#                        ranging from 0.00 to 10.00        #
#                                                          #
# VigorFinish ............. Numeric (e.g., interval)       #
#         A measure of general health and appearance,      #
#                        ranging from 0.00 to 10.00        #
############################################################
```

In an effort to promote self-documentation and readability, it is often desirable to label all object variables. First, use the epicalc::des() function and the str() function to see the nature of the data frame. Then, use the epicalc::label.var() function to provide descriptive labels for each variable. Of course, be sure to load the epicalc package, if it is not operational from prior analyses.

```
install.packages("epicalc")
library(epicalc)             # Load the epicalc package.
help(package=epicalc)        # Show the information page.
sessionInfo()                # Confirm all attached packages.

epicalc::des(LStockVg.df)
str(LStockVg.df)

epicalc::label.var(Subject,        "Animal ID",
  dataFrame=LStockVg.df)
epicalc::label.var(WeightInitial,  "Weight When Purchased",
  dataFrame=LStockVg.df)
epicalc::label.var(WeightFinish,   "Weight When Sold",
  dataFrame=LStockVg.df)
epicalc::label.var(VigorInitial,   "Vigor When Purchased",
  dataFrame=LStockVg.df)
epicalc::label.var(Vigor100Lbs,    "Vigor at 100 Pounds",
  dataFrame=LStockVg.df)
epicalc::label.var(Vigor200Lbs,    "Vigor at 200 Pounds",
  dataFrame=LStockVg.df)
epicalc::label.var(VigorFinish,    "Vigor at Sale",
  dataFrame=LStockVg.df)
```

Then confirm the description of each object variable, to be sure that all actions were deployed correctly.

```
epicalc::des(LStockVg.df)
str(LStockVg.df)
```

With assurance that the data frame is in correct format and that labels are correct, coerce objects into correct format. Once again, notice how variables are named: DataFrame$ObjectName.

```
LStockVg.df$Subject       <- as.factor(
  LStockVg.df$Subject)
LStockVg.df$WeightInitial <- as.numeric(
  LStockVg.df$WeightInitial)
LStockVg.df$WeightFinish  <- as.numeric(
  LStockVg.df$WeightFinish)
LStockVg.df$VigorInitial  <- as.numeric(
  LStockVg.df$VigorInitial)
LStockVg.df$Vigor100Lbs   <- as.numeric(
  LStockVg.df$Vigor100Lbs)
```

7.4 Conduct a Visual Data Check

```
LStockVg.df$Vigor200Lbs    <- as.numeric(
  LStockVg.df$Vigor200Lbs)
LStockVg.df$VigorFinish    <- as.numeric(
  LStockVg.df$VigorFinish)
```

As a sidebar comment, at the R prompt, key help (as.numeric) and then key help (as.integer) to see the differences between these two R functions and when it may be best to use each.

Again, confirm the structure of the dataset, using both the epicalc::des() function and the str() function.

```
epicalc::des(LStockVg.df)
str(LStockVg.df)
```

Use the summary() function against the object LStockVg.df, which is a data frame, to gain an initial sense of descriptive statistics and frequency distributions.

```
summary(LStockVg.df)
```

Now, merely use the attach() function again to assure that all data are attached to the data frame.

```
attach(LStockVg.df)
head(LStockVg.df)
tail(LStockVg.df)
summary(LStockVg.df) # Quality assurance data check

print(LStockVg.df)
```

Be sure to observe which part of the dataset shows on the screen when the print() function is used with a large dataset. The head() function and tail() function may be better alternate selections, here.

Finally, review the structure of the dataset to be assured, again, that data are in correct format.

```
str(LStockVg.df)     # List all objects, with finite detail
ls.str(LStockVg.df)  # List all objects, with finite detail
```

These many actions may be somewhat redundant, but this initial work is well-worth the effort in view of quality assurance. Every effort must be made to be sure that the data are in correct and desired format. Merely glancing at the dataset, either in an external spreadsheet or by using the print() function, is simply insufficient to meet quality assurance requirements.

7.4 Conduct a Visual Data Check

Now that the data are all in proper format, it would be common to immediately attempt the appropriate inferential analyses, Pearson's Product-Moment Coefficient of Correlation (e.g., Pearson's r) and later a linear regression for this lesson.

However, it is best to first prepare a few graphical displays of the data and to then reinforce comprehension of the data with descriptive statistics and measures of central tendency.

The value of a visual data check goes far beyond what is gained with output from the summary() function and similar functions. Simple plots can be very helpful in looking for data that may be either illogical or out-of-range. These initial plots should be considered throwaways as they are intended only for diagnostic purposes. More complex figures, often of publishable quality can then be prepared from these initial throwaway graphics, by careful selection of functions and arguments. R-based figures can be simple or detailed, as needed.

Although the emphasis in this lesson is on Pearson's r and later a linear regression, a simple graphic will be prepared for each variable, largely as a quality assurance check against the entire dataset.

```
names(LStockVg.df)      # Confirm all object variables.

par(ask=TRUE)
plot(LStockVg.df$Subject,
  main="LStockVg.df$Subject Visual Data Check")

par(ask=TRUE); plot(sort(LStockVg.df$WeightInitial))
par(ask=TRUE); plot(sort(LStockVg.df$WeightFinish))
par(ask=TRUE); plot(sort(LStockVg.df$VigorInitial))
par(ask=TRUE); plot(sort(LStockVg.df$Vigor100Lbs))
par(ask=TRUE); plot(sort(LStockVg.df$Vigor200Lbs))
par(ask=TRUE); plot(sort(LStockVg.df$VigorFinish))
  # Sorting makes it easier to look for extreme values
```

By using the ; character in the above examples, two operations (use of the par() function and use of the plot() function) could be placed on the same line. The sort() function provides a useful way of ordering data, again to look for data that may be either illogical or out-of-range.

The density plot is very useful to look for a general display of the data, largely to see distribution patterns and adherence to any semblance of normal distribution.

```
par(ask=TRUE)
plot(density(LStockVg.df$WeightInitial,
  na.rm=TRUE),    # Required for the density() function
  main="Density Plot of WeightInitial",
  lwd=6, col="red", font.axis=2, font.lab=2)

par(ask=TRUE)
plot(density(LStockVg.df$WeightFinish,
  na.rm=TRUE),    # Required for the density() function
  main="Density Plot of WeightFinish",
  lwd=6, col="red", font.axis=2, font.lab=2)
```

7.4 Conduct a Visual Data Check

```
par(ask=TRUE)
plot(density(LStockVg.df$VigorInitial,
  na.rm=TRUE),    # Required for the density() function
  main="Density Plot of VigorInitial",
  lwd=6, col="red", font.axis=2, font.lab=2)

par(ask=TRUE)
plot(density(LStockVg.df$Vigor100Lbs,
  na.rm=TRUE),    # Required for the density() function
  main="Density Plot of Vigor100Lbs",
  lwd=6, col="red", font.axis=2, font.lab=2)

par(ask=TRUE)
plot(density(LStockVg.df$Vigor200Lbs,
  na.rm=TRUE),    # Required for the density() function
  main="Density Plot of Vigor200Lbs",
  lwd=6, col="red", font.axis=2, font.lab=2)

par(ask=TRUE)
plot(density(LStockVg.df$VigorFinish,
  na.rm=TRUE),    # Required for the density() function
  main="Density Plot of VigorFinish",
  lwd=6, col="red", font.axis=2, font.lab=2)
```

The boxplot (e.g., box-and-whiskers plot) is a traditional tool for viewing the distribution of data, with an emphasis on standard descriptive statistics. See help pages for the fivenum() function and the boxplot.stats() function, along with the boxplot() function, to learn more about this visual tool.

```
par(ask=TRUE)
boxplot(LStockVg.df$WeightInitial,
  main="Boxplot of Weight When Purchased",
  col="red", lwd=2, cex.axis=1.25,
  ylab="Weight in Pounds", cex.lab=1.25)

par(ask=TRUE)
boxplot(LStockVg.df$WeightFinish,
  main="Boxplot of Weight When Sold",
  col="red", lwd=2, cex.axis=1.25,
  ylab="Weight in Pounds", cex.lab=1.25)

par(ask=TRUE)
boxplot(LStockVg.df$VigorInitial,
  main="Boxplot of Vigor When Purchased",
  col="red", lwd=2, cex.axis=1.25,
  ylab="Vigor (Scale = 0 to 10)", cex.lab=1.25)
```

```
par(ask=TRUE)
boxplot(LStockVg.df$Vigor100Lbs,
  main="Boxplot of Vigor at 100 Pounds",
  col="red", lwd=2, cex.axis=1.25,
  ylab="Vigor (Scale = 0 to 10)", cex.lab=1.25)

par(ask=TRUE)
boxplot(LStockVg.df$Vigor200Lbs,
  main="Boxplot of Vigor at 200 Pounds",
  col="red", lwd=2, cex.axis=1.25,
  ylab="Vigor (Scale = 0 to 10)", cex.lab=1.25)

par(ask=TRUE)
boxplot(LStockVg.df$VigorFinish,
  main="Boxplot of Vigor at Sale",
  col="red", lwd=2, cex.axis=1.25,
  ylab="Vigor (Scale = 0 to 10)", cex.lab=1.25)

par(ask=TRUE)
boxplot(LStockVg.df[, 2:3],      # Variables 2 to 3
  main="Comparative Boxplots of Weight",
  col="red", lwd=2, cex.axis=1.25,
  ylab="Weight in Pounds", cex.lab=1.25)

par(ask=TRUE)
boxplot(LStockVg.df[, 4:7],      # Variables 4 to 7
  main="Comparative Boxplots of Vigor",
  col="red", lwd=2, cex.axis=1.25,
  ylab="Vigor (Scale = 0 to 10)", cex.lab=1.25)
```

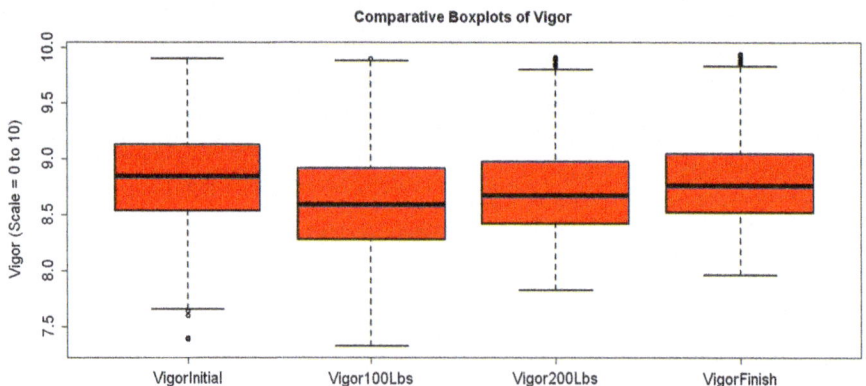

7.4 Conduct a Visual Data Check

The histogram is another traditional tool used to view data distribution. Be careful to distinguish between output of a histogram (e.g. hist() function), output of a column chart (e.g., barplot() function), and the type of data needed for each visual aid.

```
par(ask=TRUE)
hist(LStockVg.df$WeightInitial,
  main="Histogram of Weight When Purchased",
  col="red", lwd=2, cex.axis=1.25, cex.lab=1.25,
  xlab="Weight in Pounds", ylab="Number of Animals")

par(ask=TRUE)
hist(LStockVg.df$WeightFinish,
  main="Histogram of Weight When Sold",
  col="red", lwd=2, cex.axis=1.25, cex.lab=1.25,
  xlab="Weight in Pounds", ylab="Number of Animals")

par(ask=TRUE)
hist(LStockVg.df$VigorInitial,
  main="Histogram of Vigor When Purchased",
  col="red", lwd=2, cex.axis=1.25, cex.lab=1.25,
  xlab="Vigor (Scale = 0 to 10)", ylab="Number of Animals")

par(ask=TRUE)
hist(LStockVg.df$Vigor100Lbs,
  main="Histogram of Vigor at 100 Pounds",
  col="red", lwd=2, cex.axis=1.25, cex.lab=1.25,
  xlab="Vigor (Scale = 0 to 10)", ylab="Number of Animals")

par(ask=TRUE)
hist(LStockVg.df$Vigor200Lbs,
  main="Histogram of Vigor at 200 Pounds",
  col="red", lwd=2, cex.axis=1.25, cex.lab=1.25,
  xlab="Vigor (Scale = 0 to 10)", ylab="Number of Animals")

par(ask=TRUE)
hist(LStockVg.df$VigorFinish,
  main="Histogram of Vigor at Sale",
  col="red", lwd=2, cex.axis=1.25, cex.lab=1.25,
  xlab="Vigor (Scale = 0 to 10)", ylab="Number of Animals")
```

The purpose of these initial plots is to gain a general sense of the data and to equally look for outliers. In an attempt to look for outliers, the ylim argument has been avoided, so that all data are plotted. Extreme values may or may not be outliers, but they are certainly interesting and demand attention.

7.5 Descriptive Analysis of the Data

Given the different ways missing data can impact analyses, it is often helpful to first check for missing data by using the is.na() function and the complete.cases() function against the entire dataset. Both functions return a TRUE or FALSE response, depending on the function and the outcome of whether data are missing or data are not missing.

```
is.na(LStockVg.df)              # Check for missing data
complete.cases(LStockVg.df)     # Check for complete cases
```

For the dataset LStockVg.df note how there are a few rows where there are missing data for individual object variables. But, because the dataset is large (N = 3,308 subjects), everything does not always show on the screen in a convenient manner. Even so, for the purpose of this lesson, it is only necessary to recognize that there are missing data and to use appropriate functions and arguments to accommodate this observation.

Although this dataset is fairly large, the summary() function may be all that is necessary to gain a sense of the data. As typically used, the summary() function is applied against the entire dataset, thus yielding information about all object variables, including the object variable Subject.

```
summary(LStockVg.df)
```

Give attention to the listing of NAs, if any, for those object variables with missing data. Again, the summary() function is very useful and it should always be a first selection when preparing descriptive analyses.

Although the summary() function is quite sufficient, descriptive statistics for individual object variables may be desired. To achieve this aim, review the prior lesson *Descriptive Statistics and Measures of Central Tendency* for a comprehensive review of the functions used for descriptive statistics, especially: length(), asbio::Mode(), median(), mean(), sd(), table(), and finally summary(). As needed (but not always, depending on specific functions), the na.rm=TRUE argument or some other similar convention will be used to accommodate missing data.

Again, the dataset seems to be in correct form. There are no factor-type object variables in the dataset used for breakout analyses. All object variables in this dataset, with the exception of Subject, are numeric. To supplement the summary() function and determine mode, the following functions found in external packages will be used for their rich output: asbio::Mode and Zelig::Mode.

```
install.packages("asbio")
library(asbio)              # Load the asbio package.
help(package=asbio)         # Show the information page.
sessionInfo()               # Confirm all attached packages.

install.packages("Zelig")
library(Zelig)              # Load the Zelig package.
```

7.5 Descriptive Analysis of the Data

```
help(package=Zelig)      # Show the information page.
sessionInfo()            # Confirm all attached packages.
```

With these external packages brought into this R session, prepare descriptive statistics of selected variables in the LStockVg.df dataset. Use this process for all variables, as needed.

```
mean(LStockVg.df$WeightFinish, na.rm=TRUE)
sd(LStockVg.df$WeightFinish, na.rm=TRUE)
median(LStockVg.df$WeightFinish, na.rm=TRUE)
asbio::Mode(LStockVg.df$WeightFinish)
Zelig::Mode(LStockVg.df$WeightFinish)
range(LStockVg.df$WeightFinish, na.rm=TRUE) # Main descriptives
```

These one-by-one calculations of descriptive statistics are useful, but there are other functions that provide in one simple operation a composite of all needed descriptive statistics. For this lesson, look at the use of the fields::stats() function and the lessR::SummaryStats() function. Each function provides the same general level of information about the selected variables, so selection is largely a matter of preference.

```
install.packages("fields")
library(fields)             # Load the fields package.
help(package=fields)        # Show the information page.
sessionInfo()               # Confirm all attached packages.

fields::stats(LStockVg.df$WeightInitial)
fields::stats(LStockVg.df$WeightFinish)
fields::stats(LStockVg.df$VigorInitial)
fields::stats(LStockVg.df$Vigor100Lbs)
fields::stats(LStockVg.df$Vigor200Lbs)
fields::stats(LStockVg.df$VigorFinish)
fields::stats(LStockVg.df)   # Entire dataset

install.packages("lessR")
library(lessR)              # Load the lessR package.
help(package=lessR)         # Show the information page.
sessionInfo()               # Confirm all attached packages.

lessR::SummaryStats(WeightInitial, dframe=LStockVg.df)
lessR::SummaryStats(WeightFinish, dframe=LStockVg.df)
lessR::SummaryStats(VigorInitial, dframe=LStockVg.df)
lessR::SummaryStats(Vigor100Lbs,  dframe=LStockVg.df)
lessR::SummaryStats(Vigor200Lbs,  dframe=LStockVg.df)
lessR::SummaryStats(VigorFinish,  dframe=LStockVg.df)
```

7.6 Conduct the Statistical Analysis

The preceding graphical images and descriptive statistics provide a fairly good idea of the data in this dataset, dealing with the weight and vigor at various stages in the life cycle of livestock in a contemporary feeder operation:

```
Descriptive Statistics of Weight and Vigor

                    N      NAs      Mean       SD
WeightInitial    3308        0     35.01     1.99
WeightFinish     3299        9    249.86    11.84
VigorInitial     3308        0      8.83     0.42
Vigor100Lbs      3304        4      8.62     0.45
Vigor200Lbs      3294       14      8.71     0.40
VigorFinish      3299        9      8.80     0.38
```

There is every reason to think that the data are in correct format and that the data are ready for inferential analyses, such as calculation of Pearson's r and simple (e.g., one predictor variable) linear regression.

7.6.1 Correlation Using Pearson's r

The task now is to use correlation to first determine the correlation between X and Y, of which there are multiple X:Y combinations in this dataset. Only then would there be sufficient context and understanding of the data to build a regression-based prediction equation using R syntax.

R supports many possible ways to determine Pearson's r, either individual calculations, or correlation values in a table-like presentation. Different ways of showing calculation of Pearson's r are shown below, from among many different ways to achieve this aim.

Using the term *brute force*, look at the way multiple individual correlations are prepared below. View this presentation as a declining cascade of comparisons between X:Y, recognizing that an initial comparison of X:Y equates to a later comparison of Y:X (e.g, that is to say the calculation of Pearsons's r coefficient of correlation for X:Y is the same as the calculation of Pearsons's r coefficient of correlation for Y:X).

```
cor.test(LStockVg.df$WeightInitial, LStockVg.df$WeightFinish)
cor.test(LStockVg.df$WeightInitial, LStockVg.df$VigorInitial)
cor.test(LStockVg.df$WeightInitial, LStockVg.df$Vigor100Lbs)
cor.test(LStockVg.df$WeightInitial, LStockVg.df$Vigor200Lbs)
cor.test(LStockVg.df$WeightInitial, LStockVg.df$VigorFinish)
```

7.6 Conduct the Statistical Analysis

```
cor.test(LStockVg.df$WeightFinish, LStockVg.df$VigorInitial)
cor.test(LStockVg.df$WeightFinish, LStockVg.df$Vigor100Lbs)
cor.test(LStockVg.df$WeightFinish, LStockVg.df$Vigor200Lbs)
cor.test(LStockVg.df$WeightFinish, LStockVg.df$VigorFinish)

cor.test(LStockVg.df$VigorInitial, LStockVg.df$Vigor100Lbs)
cor.test(LStockVg.df$VigorInitial, LStockVg.df$Vigor200Lbs)
cor.test(LStockVg.df$VigorInitial, LStockVg.df$VigorFinish)

cor.test(LStockVg.df$Vigor100Lbs, LStockVg.df$Vigor200Lbs)
cor.test(LStockVg.df$Vigor100Lbs, LStockVg.df$VigorFinish)

cor.test(LStockVg.df$Vigor200Lbs, LStockVg.df$VigorFinish)
```

The cor.test() function is quite descriptive, yielding information on the p-value and Pearson's r coefficient of correlation (Pearson's r is the default method for correlation when using the cor.test() function). Be careful to distinguish between p-values and Pearson's r correlations, however. As an example, look at the output for the cor.test() function when applied against VigorInitial and VigorFinish:

```
> cor.test(LStockVg.df$VigorInitial, LStockVg.df$VigorFinish)

        Pearson's product-moment correlation

data:   LStockVg.df$VigorInitial and LStockVg.df$VigorFinish
t = 20.1132, df = 3297, p-value < 0.00000000000000022
alternative hypothesis: true correlation is not equal to 0
95 percent confidence interval:
 0.2998468 0.3606473
sample estimates:
     cor
0.33059
>
```

The calculated p-value is less than or equal to 0.00000000000000022 and as such there is a statistically significant correlation (p <= 0.05) between VigorInitial and VigorFinish. All we know from the p-value statistic is that there is (or is not) a statistically significant correlation, and in this example there is a significant correlation.

The Pearson's r coefficient of correlation is 0.33059. Recall that Pearson's r is the default method for determining correlation with the cor.test() function.

It is now known that the correlation (Pearons's r) between VigorInitial and VigorFinish is 0.33059 and that this correlation is statistically significant (p <= 0.05). That is to say, as the measure for VigorInitial increases there is a corresponding increase in the measure for VigorFinish. It should be noted, however, that coefficients of correlation are very sensitive to N and this should be recalled when considering the practical implications of these analyses.

Although these multiple individual correlations are adequate, there are a few simple ways to avoid all of this redundant typing and instead use R-based functions that simplify the process. Equally, the output to the screen of these simplified operations is generally easy to read and does not call for scrolling up and down the screen to find desired statistics.

First, because it is not necessary, remove the Subject column from the data frame so that only the data pertinent (e.g., the measured object variables, the two measures for weight and the four measures for vigor in this lesson) to this set of correlations are retained. Although this action is not essential, it will make implementation of the analyses a bit easier than if the LStockVg.df$Subject object variable were retained.

```
str(LStockVg.df)
head(LStockVg.df)

LStockVg.df$Subject <- NULL  # Remove the Subject column

attach(LStockVg.df)           # Confirm data are attached

str(LStockVg.df)
head(LStockVg.df)
```

Now that the revised LStockVg.df data frame is in corrected format, with Subject removed, apply the cor() function and Hmisc::rcor() function to generate attractive and easy-to-read correlation output.

The cor() function produces a convenient correlation matrix of Pearson's r values. Be sure to notice how the X:X, Y:Y, Z:Z, etc. correlations always show as 1.00 and appear as a sloping diagonal, from top left to bottom right. This type of output is common to any correlation matrix and it is by no means unique to R.

```
cor(LStockVg.df, use="complete.obs", method="pearson")
```

The Hmisc::rcorr() function is information-rich and it generates three types of output: Pearson's r values for each comparison of X:Y, N for each X:Y comparison with missing values taken out of the comparisons, and X:Y p-values.

```
install.packages("Hmisc")
library(Hmisc)              # Load the Hmisc package.
help(package=Hmisc)         # Show the information page.
sessionInfo()               # Confirm all attached packages.

Hmisc::rcorr(as.matrix(LStockVg.df, type=pearson))
```

Note above how the data frame must be accommodated as a matrix for the Hmisc::rcorr() function to work. This requirement is easily accommodated by wrapping the as.matrix() function around the data frame name, as shown above.

The placement of Pearson's r correlation values, Ns, and p-values in an organized fashion is highly valued, but going back to a constant theme in these lessons on the use of R for biostatistics, visual presentations are perhaps the best way to gain attention of the typical reader. Consider the tools shown below on how

7.6 Conduct the Statistical Analysis

to present multiple correlation comparisons in one convenient image, using the pairs() function, the lattice::splom() function, the psych::pairs.panels() function, the car::scatterplotMatrix() function, and the psych::cor.plot() function. Each function addresses the same general theme, but presentation is slightly different, allowing a variety of selections for final presentation.

The pairs() function is likely the first choice for production of a visual correlation matrix. Of course, the visual representation of X:Y is very helpful as a supplement to the otherwise static Pearson's r coefficient statistics.

```
par(ask=TRUE)
pairs(~WeightInitial+WeightFinish+VigorInitial+Vigor100Lbs+
   Vigor200Lbs+VigorFinish, data=LStockVg.df, col="red",
   main="Scatter Plot Matrix (SPLOM) of Weights and Vigor
   in a Livestock Finishing Operation")
```

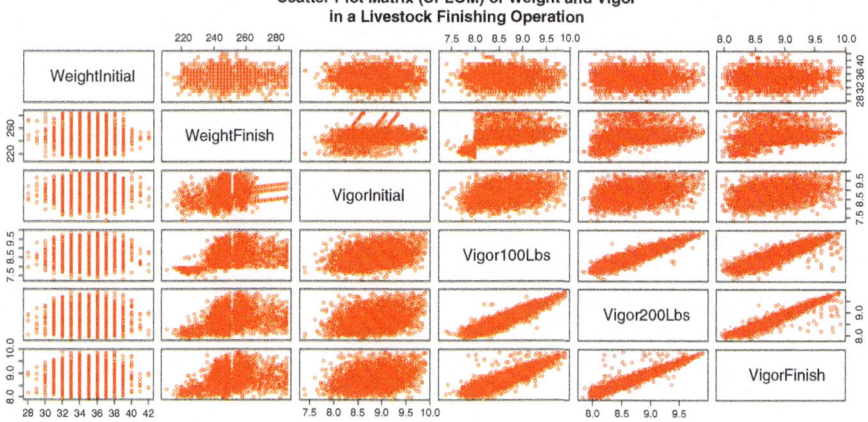

The lattice::splom() function is also useful, but be sure to see how the presentation improves as the number of comparisons is set to a limited number, or in the examples below as the number of comparisons is reduced from six to three.

```
install.packages("lattice")
library(lattice)            # Load the lattice package.
help(package=lattice)       # Show the information page.
sessionInfo()               # Confirm all attached packages.

par(ask=TRUE)
lattice::splom(~LStockVg.df, font=2, col="red",
   main="Scatter Plot Matrix (SPLOM) of Weights and Vigor
   in a Livestock Finishing Operation")
   # All 6 variables for this splom
```

```
par(ask=TRUE)
lattice::splom(~LStockVg.df[4:6], font=2, col="red",
  main="Scatter Plot Matrix (SPLOM) of Vigor in a
  Livestock Finishing Operation")
# Look at the way only variables 4, 5, and 6 are used
```

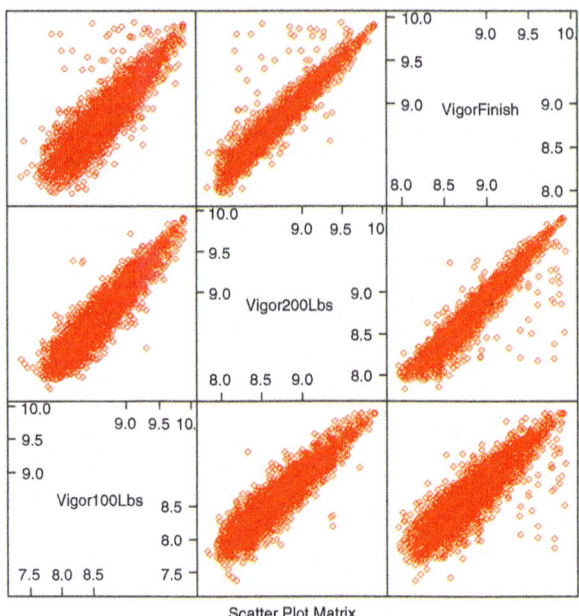

Scatter Plot Matrix

Following along with the different ways R supports the creation of a scatter plot matrix (SPLOM), consider use of the psych::pairs.panels() function. The output is as appealing as the other SPLOM-type functions and it may be easier to read. Be sure to notice how the output includes a histogram and density plot of each individual variable and then a scatter plot of each X:Y comparison.

```
install.packages("psych")
library(psych)              # Load the psych package.
help(package=psych)         # Show the information page.
sessionInfo()               # Confirm all attached packages.

par(ask=TRUE)
psych::pairs.panels(LStockVg.df,
  method="pearson", rug=TRUE, hist.col="red",
  main="Scatter Plot Matrix (SPLOM) (Lower Diagonal) and
  Pearson's r (Upper Diagonal) of Weight and Vigor")
```

7.6 Conduct the Statistical Analysis

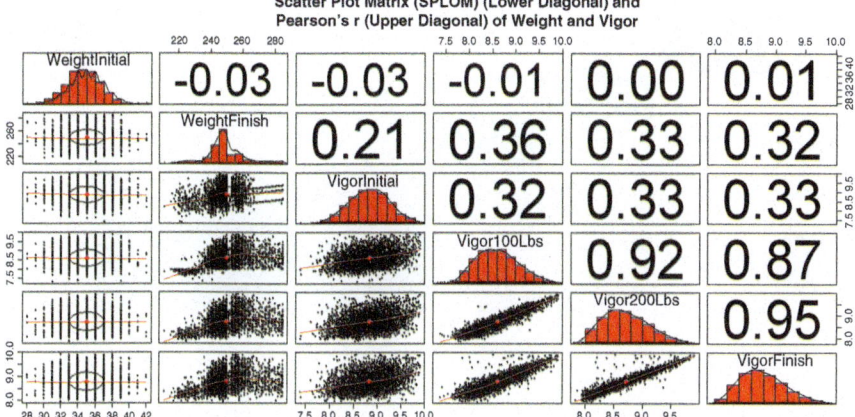

The car::scatterplotMatrix() function provides another view of a scatter plot matrix. Again, there are many options and eventually individual preferences often determine selected functions.

```
install.packages("car")
library(car)                  # Load the car package.
help(package=car)             # Show the information page.
sessionInfo()                 # Confirm all attached packages.

car::scatterplotMatrix(~ VigorInitial + Vigor100Lbs +
  Vigor200Lbs + VigorFinish,    # Only vigor indicators
  main="Scatterplot Matrix of All Vigor Indicators",
  transform=TRUE, data=LStockVg.df, smoother=loessLine,
  legend.plot=TRUE, row1attop=TRUE)
```

As an interesting change from the traditional SPLOM, look at output from the psych::cor.plot() function and how color gradients are used to signify the degree of correlation (−1.0 to +1.0) and subsequently Pearson's r, ranging from −1.0 (Dark Red) to 0.00 (White) to +1.00 (Dark Blue).

```
par(ask=TRUE)
psych::cor.plot(cor(LStockVg.df,
  use="complete.obs", method="pearson"),
  main="Color-Gradient Correlation Plot of Weight and Vigor:
  Dark Red for Pearson's r = -1.0 to Dark Blue for
  Pearson's r = +1.0", font.lab=2, font.axis=2)
```

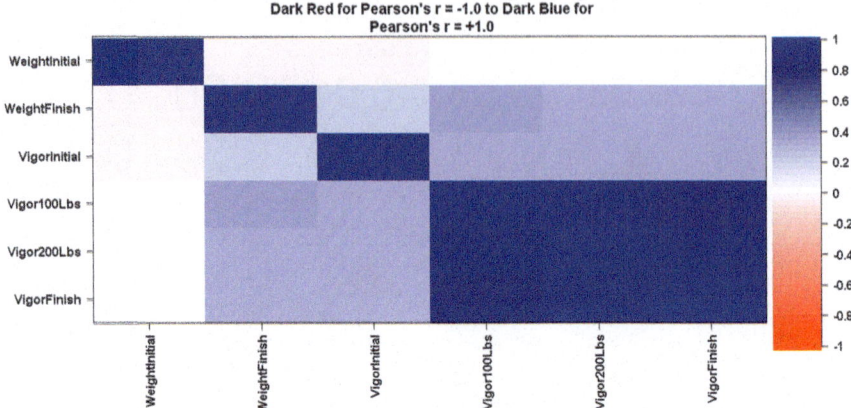

If individual scatter plots of X:Y are needed, the car package has a few useful functions. A regression line can be inserted into a figure generated from the plot() function by using the car::regLine() function. Or, simply use the car::scatterplot() function to generate an individual scatter plot with accompanying regression lines. In the examples shown below, from among the many X:Y comparisons, look at the two ways a regression line has been added to the presentation, for VigorFinish (Y axis) by Vigor200Lbs (X axis) in this example.

```
savelwd       <- par(lwd=4)           # Heavy line
savefont      <- par(font=2)          # Bold
savecex.lab   <- par(cex.lab=1.25)    # Label
savecex.axis  <- par(cex.axis=1.25)   # Axis
par(ask=TRUE)
plot(VigorFinish ~ Vigor200Lbs,       # Y axis ~ X axis
  data=LStockVg.df,
  xlab="Vigor at 200 Pounds", ylab="Vigor at Sale",
  main="Scatter Plot of Vigor at 200 Pounds (X) by Vigor at
  Sale (Y) With Regression Line",
  pch=c(16))
car::regLine(lm(VigorFinish ~ Vigor200Lbs,
  data=LStockVg.df), lty=1, col="red")
par(savelwd); par(savefont); par(savecex.lab);
par(savecex.axis)

savelwd       <- par(lwd=4)           # Heavy line
savefont      <- par(font=2)          # Bold
savecex.lab   <- par(cex.lab=1.5)     # Label
savecex.axis  <- par(cex.axis=1.5)    # Axis
par(ask=TRUE)
car::scatterplot(VigorFinish ~ Vigor200Lbs,    # Y ~ X
  data=LStockVg.df,
  xlab="Vigor at 200 Pounds", ylab="Vigor at Sale",
```

7.6 Conduct the Statistical Analysis

```
  main="Scatter Plot of Vigor at 200 Pounds (X) by Vigor at
  Sale (Y) With Regression Line and Ellipse",
  reg.line=TRUE, boxplots="xy", grid=FALSE, pch=15,
  font.lab=2, font.axis=2, lty=1, cex.main=1.5,
  ellipse=TRUE, robust=TRUE )
  # Data-concentration ellipse and center of ellipse
legend("topleft",
  xjust=1, bty="y", box.lwd=1, box.col="red",
  text.col="red", text.font=2,
  "           Pearson's r = 0.9466428    ")
par(savelwd); par(savefont); par(savecex.lab)
par(savecex.axis)
```

The X:Y scatter plot is well-established in biostatistics and there are many ways to prepare and present a scatter plot. A fairly new approach, however, is to use the bagplot to show the degree of association between two separate variables, X:Y. The bagplot is a bivariate boxplot, where 50 % of all points are contained in the central bag. A fence surrounds the bag and from this fence the remaining points radiate out, giving a view of distribution points and consequently extreme values if there are any. Compare the visualization of the association between Vigor200Lbs (X axis) and VigorFinish (Y axis) in a bagplot (below) as compared to the more traditional X:Y scatter plot of these two object variables, as shown above.

```
install.packages("aplpack")
library(aplpack)              # Load the aplpack package.
help(package=aplpack)         # Show the information page.
sessionInfo()                 # Confirm all attached packages.

savelwd      <- par(lwd=2)          # Heavy line
savefont     <- par(font=2)         # Bold
savecex.lab  <- par(cex.lab=1.5)    # Label
savecex.axis <- par(cex.axis=1.5)   # Axis
par(ask=TRUE)
aplpack::bagplot(LStockVg.df$Vigor200Lbs,
  LStockVg.df$VigorFinish,
  main="Bagplot of Vigor at 200 Pounds (X) by Vigor at Sale (Y)
  With Central Bag, Fence, and Distribution Points",
  na.rm=TRUE,                 # Accommodate missing data
  xlab="Vigor at 200 Pounds", ylab="Vigor at Sale",
  show.outlier=TRUE,          # At the R prompt, key )
  show.whiskers=TRUE,         # help(bagplot) to see details for
  show.looppoints=TRUE,       # each argument, show.outlier,
  show.bagpoints=TRUE,        # show.whiskers,etc. Then, decide
  show.loophull=TRUE,         # which arguments meet individual
  show.baghull=TRUE,          # needs.
  pch=c(22))                  # Filled square red symbol
par(savelwd); par(savefont); par(savecex.lab)
par(savecex.axis)
```

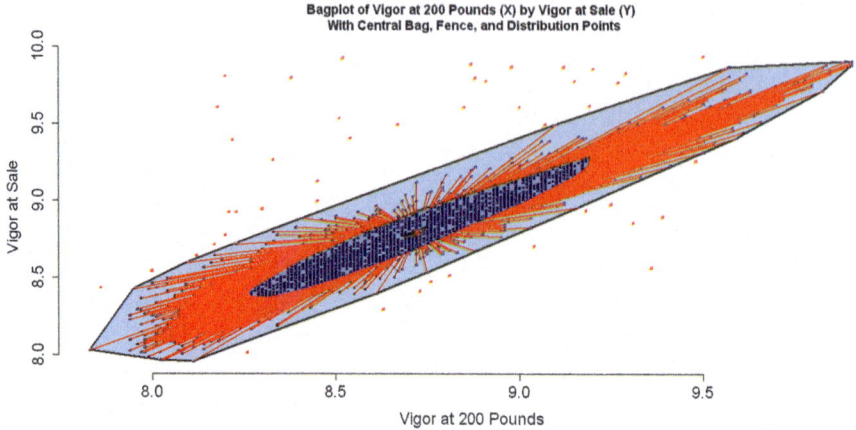

7.6.2 Linear Regression

For this lesson, the main interest in terms of importance should be directed at the object variable WeightFinish. Profits for this livestock finishing operation are made or lost based on WeightFinish: how quickly desired (in this lesson, 250 pounds) WeightFinish is achieved, the amount and cost of management and feed needed to reach desired WeightFinish, etc. Given this interest in WeightFinish, the remaining task for this lesson is to prepare a prediction equation for those object variables that show significant (and practical) correlation with WeightFinish: VigorInitial, Vigor100Lbs, Vigor200Lbs, VigorFinish. Although it is beyond the immediate purpose of this lesson, it might be helpful to also look into the issue of multicollinearity, but again, multicollinearity is beyond the purpose of this introductory lesson.

To achieve this aim, prepare a prediction equation (solve for Y-hat, or $\hat{Y}$ to be precise when using an editor that supports specialized symbols) for each individual variable of interest (VigorInitial, Vigor100Lbs, Vigor200Lbs, VigorFinish) against WeightFinish. For each prediction equation, the focus is on a simple linear regression, where one and only one object variable is put into the prediction equation.

Prediction of WeightFinish Using VigorInitial

```
Fit.WeightFinish.by.VigorInitial <- lm(WeightFinish ~ 
  VigorInitial, data=LStockVg.df)

anova(Fit.WeightFinish.by.VigorInitial)
  # Confirm significance
```

7.6 Conduct the Statistical Analysis

```
Response: WeightFinish
              Df  Sum Sq  Mean Sq  F value    Pr(>F)
VigorInitial   1   21214  21214.4   158.43  < 2.2e-16 ***
Residuals   3297  441471    133.9

summary(Fit.WeightFinish.by.VigorInitial)
  # Obtain statistics needed for prediction equation

Coefficients:
              Estimate  Std. Error  t value  Pr(>|t|)
(Intercept)   196.4606      4.2470    46.26    <2e-16 ***
VigorInitial    6.0438      0.4802    12.59    <2e-16 ***
```

Based on this output, there is certainly significance (p <= 0.05) and the prediction equation is:

```
Y-hat = a + b(x)

WeightFinish = 196.4606 + VigorInitial(6.0438)
```

Assume that an individual animal had a VigorInitial value of 8.85. What is the predicted value for WeightFinish? Apply the above prediction equation:

```
WeightFinish = 196.4606 + (8.85 * 6.0438)
WeightFinish = 196.4606 + 53.48763
WeightFinish = 249.94823
```

The livestock manager now knows that for this large-scale feeder operation, an animal with a VigorInitial value of approximately 8.85 will yield a finishing weight (WeightFinish) of approximately 250 pounds.

As an interesting complement to this part of the lesson, apply the predict() function against the model to see possibilities at the level of individual subjects:

```
predict(Fit.WeightFinish.by.VigorInitial, interval= "prediction")
  # Be sure to view the Warning message on future responses
```

Prediction of WeightFinish Using Vigor100Lbs

```
Fit.WeightFinish.by.Vigor100Lbs <- lm(WeightFinish ~
  Vigor100Lbs, data=LStockVg.df)

anova(Fit.WeightFinish.by.Vigor100Lbs)
  # Confirm significance
```

```
Response: WeightFinish
              Df  Sum Sq  Mean Sq  F value    Pr(>F)
Vigor100Lbs    1   59658    59658   488.04  < 2.2e-16 ***
Residuals   3297  403027      122
```

```
summary(Fit.WeightFinish.by.Vigor100Lbs)
  # Obtain statistics needed for prediction equaltion
```

```
Coefficients:
             Estimate Std. Error t value Pr(>|t|)
(Intercept)  168.5647     3.6849   45.74   <2e-16 ***
Vigor100Lbs    9.4289     0.4268   22.09   <2e-16 ***
```

Based on this output, there is certainly significance ($p <= 0.05$) and the prediction equation is:

```
Y-hat = a + b(x)
```

```
WeightFinish = 168.5647 + Vigor100Lbs(9.4289)
```

Assume that an individual animal had a Vigor100Lbs value of 8.63. What is the predicted value for WeightFinish? Apply the above prediction equation:

```
WeightFinish = 168.5647 + (8.63 * 9.4289)
WeightFinish = 168.5647 + 81.371407
WeightFinish = 249.936107
```

The livestock manager now knows that for this large-scale feeder operation, an animal with a Vigor100Lbs value of approximately 8.63 will yield a finishing weight (WeightFinish) of approximately 250 pounds.

As an interesting complement to this part of the lesson, apply the predict() function against the model to see possibilities at the level of individual subjects:

```
predict(Fit.WeightFinish.by.Vigor100Lbs, interval= "prediction")
  # Be sure to view the Warning message on future responses
```

Prediction of WeightFinish Using Vigor200Lbs

```
Fit.WeightFinish.by.Vigor200Lbs <- lm(WeightFinish ~
    Vigor200Lbs, data=LStockVg.df)
```

```
anova(Fit.WeightFinish.by.Vigor200Lbs)
  # Confirm significance
```

```
Response: WeightFinish
              Df  Sum Sq  Mean Sq  F value    Pr(>F)
```

7.6 Conduct the Statistical Analysis

```
Vigor200Lbs      1   49547    49547   396.04  < 2.2e-16 ***
Residuals     3291  411730      125

summary(Fit.WeightFinish.by.Vigor200Lbs)
  # Obtain statistics needed for prediction equaltion

Coefficients:
              Estimate Std. Error t value Pr(>|t|)
(Intercept)   164.7449     4.2814   38.48   <2e-16 ***
Vigor200Lbs     9.7679     0.4908   19.90   <2e-16 ***
```

Based on this output, there is certainly significance (p <= 0.05) and the prediction equation is:

```
Y-hat = a + b(x)

WeightFinish = 164.7449 + Vigor200Lbs(9.7679)
```

Assume that an individual animal had a Vigor200Lbs value of 8.04. What is the predicted value for WeightFinish? Apply the above prediction equation:

```
# WeightFinish = 164.7449 + (8.04 * 9.7679)
# WeightFinish = 164.7449 + 78.533916
# WeightFinish = 243.278816
```

The livestock manager now knows that for this large-scale feeder operation, an animal with a Vigor200Lbs value of approximately 8.04 will yield a finishing weight (WeightFinish) of approximately 243 pounds, which is less than the desired weight of 250 pounds and this discrepancy may greatly impact profits.

As an interesting complement to this part of the lesson, apply the predict() function against the model to see possibilities at the level of individual subjects:

```
predict(Fit.WeightFinish.by.Vigor200Lbs, interval= "prediction")
  # Be sure to view the Warning message on future responses
```

Prediction of WeightFinish Using VigorFinish

```
Fit.WeightFinish.by.VigorFinish <- lm(WeightFinish ~
  VigorFinish, data=LStockVg.df)

anova(Fit.WeightFinish.by.VigorFinish)
  # Confirm significance

Response: WeightFinish
             Df Sum Sq Mean Sq F value     Pr(>F)
```

```
VigorFinish     1   46444   46444  367.88 < 2.2e-16 ***
Residuals    3297  416241     126
```

```
summary(Fit.WeightFinish.by.VigorFinish)
  # Obtain statistics needed for prediction equaltion
```

```
Coefficients:
            Estimate Std. Error t value Pr(>|t|)
(Intercept)  162.575      4.555   35.69   <2e-16 ***
VigorFinish    9.916      0.517   19.18   <2e-16 ***
```

Based on this output, there is certainly significance (p <= 0.05) and the prediction equation is:

```
Y-hat = a + b(x)

WeightFinish = 162.575 + VigorFinish(9.916)
```

Assume that an individual animal had a VigorFinish value of 9.06. What is the predicted value for WeightFinish? Apply the above prediction equation:

```
# WeightFinish = 162.575 + (9.06 * 9.916)
# WeightFinish = 162.575 + 89.83896
# WeightFinish = 252.41396
```

The livestock manager now knows that for this large-scale feeder operation, an animal with a VigorFinish value of approximately 9.06 will yield a finishing weight (WeightFinish) of approximately 252 pounds.

As an interesting complement to this part of the lesson, apply the predict() function against the model to see possibilities at the level of individual subjects:

```
predict(Fit.WeightFinish.by.VigorFinish, interval= "prediction")
  # Be sure to view the Warning message on future responses
```

7.7 Summary

The dataset for this lesson was selected because it provided a demonstration of how researchers need to consider significant correlations but also practical correlations. The graphics and statistics provided a great deal of basic information. However, focus on outcomes of the many X:Y correlations (based on Pearson's r) and the few linear regressions which were provided to gain a sense of prediction.

As an example, look at the correlation between WeightFinish and VigorInitial. The Pearson's r coefficient of correlation for this comparison is 0.2143682 and the correlation between the two variables is statistically significant (p <= 0.05). It may not seem overly dramatic to have a Pearson's r of 0.2143682 when looking at other correlations such as the Pearson's r for Vigor200Lbs and VigorFinish, which exceeds 0.95.

However, correlation is very sensitive to N and for this large livestock finishing operation of more than 3,300 animals a statistically significant correlation that also has a Pearson's r of 0.2143682 simply cannot be ignored. Given the economy of scale gained with large operations, the operator has the potential to greatly increase profits even by only small changes in management practices when there is a Pearson's r of 0.2143682 for selected variables and conditions that impact those variables, especially with our increasing interest in data in the large or *Big Data*. Even a small increase in VigorInitial can greatly impact profits given the association between VigorInitial and WeightFinish. Does the same finding apply to WeightFinish and the three other vigor-related variables: Vigor100Lbs, Vigor200Lbs, and VigorFinish?

This lesson also dealt with the development of a prediction equation, or the attempt to use currently known values to solve for a future variable. As stated earlier, past behavior is the best predictor of future behavior and regression-based prediction equations are based on this concept.

This lesson focused on the use of one variable to predict the future value of another variable, or in this lesson the use of vigor-type values to predict finished weight. A manual effort was used for this simple prediction equation and this lesson will introduce the manual introduction of two predictor variables in the Addendum. More automated practices at prediction are also introduced in the Addendum.

The constant reminder of correlation and regression is the need to avoid any attempt to apply causation. That is to say, X may be correlated (e.g., associated) with Y, but that does not mean that X causes Y. Consider the well-accepted observation that family income is a predictor of grade point average (GPA) in the first year of college. That does not mean that a rise in family income immediately results in an increase in GPA and this is certainly not a guaranteed outcome at the level of individual subjects. Family income does not cause GPA. Performance on exams causes GPA. Instead, the two variables (family income and GPA) are correlated to each other. This example from the social sciences should be recalled when applying the use of correlation and regression to biostatistics. Causation cannot be assumed even when selected variables are used to build a future prediction equation.

7.8 Addendum: Multiple Regression

From among the many variables used in this lesson, never forget that the finishing operator responsible for the livestock needs to be practical. Consider the four measures for vigor: VigorInitial, Vigor100Lbs, Vigor200Lbs, VigorFinish. The desired weight at time of sale, when the livestock are taken to slaughter, is about 250 pounds. The manager has the best opportunity to influence finished weight by applying sound practices at the beginning (when VigorInitial is measured) and early on (when Vigor100Lbs is measured). Management options that influence future finished weight at 250 pounds are limited when Vigor200Lbs is measured and of course practices are even more limited when VigorFinish is measured.

7.8.1 Hand-Calculate Multiple Regression

Given the importance of early intervention, construct a model for VigorEarly, which takes into account management practices up to the time Vigor100Lbs is measured. Look at how two (or more) variables can be manually entered into the construction of a prediction equation for Y-hat, $\hat{Y}$.

```
Fit.WeightFinish.by.VigorEarly <- lm(WeightFinish ~
   VigorInitial + Vigor100Lbs,
   data=LStockVg.df)
   # Note how this model includes two predictors and
   # enumeration of a new concept, VigorEarly

anova(Fit.WeightFinish.by.VigorEarly)
   # Confirm significance

Response: WeightFinish
              Df  Sum Sq  Mean Sq  F value     Pr(>F)
VigorInitial   1   21214    21214   175.67 < 2.2e-16 ***
Vigor100Lbs    1   43447    43447   359.78 < 2.2e-16 ***
Residuals   3296  398024      121

summary(Fit.WeightFinish.by.VigorEarly)   # Prediction equation
   # Obtain statistics needed for prediction equaltion

Coefficients:
              Estimate Std. Error t value Pr(>|t|)
(Intercept)   149.1843     4.7412  31.465  < 2e-16 ***
VigorInitial    3.1004     0.4817   6.437 1.4e-10 ***
Vigor100Lbs     8.4996     0.4481  18.968  < 2e-16 ***
```

Before $\hat{Y}$, for what is now a multiple regression, is calculated, apply the coefficients() function and the na.action() function against the enumerated lm-type object Fit.WeightFinish.by.VigorEarly, to gain a better sense of the model.

```
coefficients(Fit.WeightFinish.by.VigorEarly)
na.action(Fit.WeightFinish.by.VigorEarly)
```

Based on this output, there is certainly significance (p <= 0.05) and the prediction equation is:

```
Y-hat = a + b(x) + b(y)

WeightFinish = 149.1843 + VigorInitial(3.1004) + Vigor100Lbs(8.4996)
```

7.8 Addendum: Multiple Regression

Assume that an individual animal had a VigorInitial value of 8.73 and a Vigor100Lbs vale of 8.63. What is the predicted value for WeightFinish? Apply the above prediction equation:

```
WeightFinish = 149.1843 + (8.73 * 3.1004) + (8.63 * 8.4996)
WeightFinish = 149.1843 + 27.066492 + 73.351548
WeightFinish = 249.60234
```

The livestock manager now knows that for this large-scale feeder operation, an animal with a VigorInitial value of 8.73 and a Vigor100Lbs vale of 8.63 will yield a finishing weight (WeightFinish) of approximately 250 pounds.

Although it may be a bit redundant, apply the cor.test() function against VigorInitial and Vigor100Lbs to see the relationship between these two variables and then reinforce the outcome with a visual image of the correlation. A plot would also help establish the relationship.

```
cor.test(LStockVg.df$VigorInitial, LStockVg.df$Vigor100Lbs)

savefont     <- par(font=2)          # Bold
par(ask=TRUE)
plot(LStockVg.df$VigorInitial, LStockVg.df$Vigor100Lbs,
  xlab="Vigor at Purchase (Approximately 35 Pounds)",
  ylab="Vigor at 100 Pounds",
  font.lab=2, font.axis=2, cex.lab=1.5, cex.main=1.5,
  pch=c(15), # square block
  main="Scatter Plot of Vigor at Purchase (X) by Vigor at
  100 Pounds (Y) with Added Regression Line")
abline(lm(Vigor100Lbs ~ VigorInitial, data =LStockVg.df),
  lwd=6, col="red")
legend("bottomleft",
   bty="n", "p-value <= 0.00000000000000022")
legend("bottomright",
   bty="n", "Pearson's r = 0.9466428              ")
par(savefont)
```

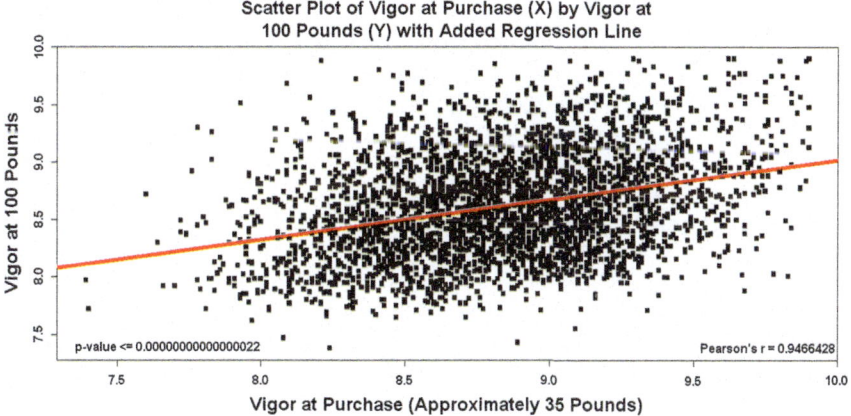

7.8.2 Minimal Adequate Model (MAM) for Regression

Multiple regression is not restricted to the use of only two predictors (VigorInitial and Vigor100Lbs, shown above) and it equally does not need to be a hand-calculation. Below, notice how all four indicators of vigor (VigorInitial, Vigor100Lbs, Vigor200Lbs, and VigorFinish) are automatically introduced into a prediction equation. A Minimal Adequate Model (MAM) approach will be used for this approach at multiple regression.

```
Fit.Model.WeightFinish.Vigor <- lm(WeightFinish ~
  VigorInitial + Vigor100Lbs + Vigor200Lbs + VigorFinish,
  data=LStockVg.df)

summary(Fit.Model.WeightFinish.Vigor)

Coefficients:
             Estimate Std. Error t value Pr(>|t|)
(Intercept)  149.9997     5.2669  28.480  < 2e-16 ***
VigorInitial   3.1481     0.4849   6.492 9.72e-11 ***
Vigor100Lbs    9.5630     1.1021   8.677  < 2e-16 ***
Vigor200Lbs   -2.6569     1.8965  -1.401    0.161
VigorFinish    1.4491     1.5721   0.922    0.357

summary.aov(Fit.Model.WeightFinish.Vigor)

               Df  Sum Sq Mean Sq F value Pr(>F)
VigorInitial    1   21197   21197  175.67 <2e-16 ***
Vigor100Lbs     1   43088   43088  357.08 <2e-16 ***
Vigor200Lbs     1     139     139    1.15  0.284
VigorFinish     1     103     103    0.85  0.357
Residuals    3288  396751     121
```

The p-value for VigorFinish exceeds 0.05 and is the greatest of the two object variables that exceed 0.05. Remove VigorFinish from the model, to see the effect of this action on model building and then attempt the same set of actions, but now with the selected object variable removed from the model.

```
Fit.Model.WeightFinish.Vigor2 <- update(
  Fit.Model.WeightFinish.Vigor, .~.-VigorFinish,
  data=LStockVg.df)
```

Note the . character that goes before the ~ character and the . character that comes after the ~ character. In this case, the . character means *same*. The important thing to recall is that VigorFinish is removed from the model, by placing a - character in front of the object variable to be removed.

7.8 Addendum: Multiple Regression

```
summary(Fit.Model.WeightFinish.Vigor2)
```

Coefficients:

	Estimate	Std. Error	t value	Pr(>|t|)	
(Intercept)	151.2207	5.0975	29.666	< 2e-16	***
VigorInitial	3.1721	0.4842	6.552	6.58e-11	***
Vigor100Lbs	9.5520	1.1020	8.668	< 2e-16	***
Vigor200Lbs	-1.3466	1.2555	-1.073	0.284	

```
summary.aov(Fit.Model.WeightFinish.Vigor2)
```

	Df	Sum Sq	Mean Sq	F value	Pr(>F)	
VigorInitial	1	21197	21197	175.68	<2e-16	***
Vigor100Lbs	1	43088	43088	357.10	<2e-16	***
Vigor200Lbs	1	139	139	1.15	0.284	
Residuals	3289	396853	121			

Vigor200Lbs remains at a p-value level greater than 0.05. Remove the remaining object variable for which there is no significance (p <= 0.05), or Vigor200Lbs in this example.

```
Fit.Model.WeightFinish.Vigor3 <- update(
  Fit.Model.WeightFinish.Vigor2, .~.-Vigor200Lbs,
  data=LStockVg.df)

summary(Fit.Model.WeightFinish.Vigor3)
```

Coefficients:

	Estimate	Std. Error	t value	Pr(>|t|)	
(Intercept)	149.1843	4.7412	31.465	< 2e-16	***
VigorInitial	3.1004	0.4817	6.437	1.4e-10	***
Vigor100Lbs	8.4996	0.4481	18.968	< 2e-16	***

```
summary.aov(Fit.Model.WeightFinish.Vigor3)
```

	Df	Sum Sq	Mean Sq	F value	Pr(>F)	
VigorInitial	1	21214	21214	175.7	<2e-16	***
Vigor100Lbs	1	43447	43447	359.8	<2e-16	***
Residuals	3296	398024	121			

And now all remaining predictors (VigorInitial and Vigor100Lbs) are significant. Use these predictors, as needed, for practical considerations in development of a prediction equation and possible attention to management practices. Biostatistics is not divorced from economics and in many situations the two can easily complement each other.

7.8.3 Stepwise Regression

As always with R, there is generally more than one way of attempting an analysis. In the below example of multiple regression note how a stepwise regression is used, as opposed to the earlier Minimal Adequate Model (MAM).

Because missing values impact stepwise regression and because there are only a few missing values in this dataset of 3,308 subjects, use the na.omit() function to remove all cases with missing data.

```
str(LStockVg.df)
summary(LStockVg.df)

LStockVg.df <- na.omit(LStockVg.df)
  # Remove all cases with missing values

str(LStockVg.df)
summary(LStockVg.df)
```

Note the change from 3,308 observations of six variables to 3,293 observations of six variables.

With the dataset now in correct format, apply the step() function to conduct the stepwise regression. Again, it is beyond the purpose of this introductory lesson on R to go into too much detail on the theory of regression and differences in methods, such as the difference (if any) between a forward stepwise regression, backward stepwise regression, etc. Fortunately, there are no limits to the available resources. Take advantage of these resources.

Construct the object Fit.Model.WeightFinish.Vigor again, incorporating all four vigor measures. Recall that as opposed to the prior construction of this model, all cases with missing values have been removed but of course there were only a few cases with missing data.

```
Fit.Model.WeightFinish.Vigor <- lm(WeightFinish ~
  VigorInitial + Vigor100Lbs + Vigor200Lbs + VigorFinish,
  data=LStockVg.df)
  # Reminder:  all cases with missing data were removed

summary(Fit.Model.WeightFinish.Vigor)

Coefficients:
              Estimate Std. Error t value Pr(>|t|)
(Intercept)   149.9997     5.2669  28.480  < 2e-16 ***
VigorInitial    3.1481     0.4849   6.492 9.72e-11 ***
Vigor100Lbs     9.5630     1.1021   8.677  < 2e-16 ***
Vigor200Lbs    -2.6569     1.8965  -1.401    0.161
VigorFinish     1.4491     1.5721   0.922    0.357
```

7.8 Addendum: Multiple Regression

```
summary.aov(Fit.Model.WeightFinish.Vigor)

              Df  Sum Sq  Mean Sq  F value  Pr(>F)
VigorInitial   1   21197    21197   175.67  <2e-16 ***
Vigor100Lbs    1   43088    43088   357.08  <2e-16 ***
Vigor200Lbs    1     139      139     1.15   0.284
VigorFinish    1     103      103     0.85   0.357
Residuals   3288  396751      121
```

Apply the step() function to begin the stepwise regression process. For this example, use a backward stepwise regression. Notice, below, how the model steps (in a backward fashion) until only those object variables of importance to the construction of a prediction equation remain. The coefficients are then shown for those remaining variables, VigorInitial and Vigor100Lbs in this example.

```
step(Fit.Model.WeightFinish.Vigor, direction="backward")

Start:  AIC=15788.44
WeightFinish ~ VigorInitial + Vigor100Lbs + Vigor200Lbs + VigorFinish

               Df  Sum of Sq    RSS    AIC
- VigorFinish   1      102.5 396853  15787
- Vigor200Lbs   1      236.8 396988  15788
<none>                       396751  15788
- VigorInitial  1     5086.3 401837  15828
- Vigor100Lbs   1     9086.0 405837  15861

Step:  AIC=15787.29
WeightFinish ~ VigorInitial + Vigor100Lbs + Vigor200Lbs

               Df  Sum of Sq    RSS    AIC
- Vigor200Lbs   1      138.8 396992  15786
<none>                       396853  15787
- VigorInitial  1     5179.1 402033  15828
- Vigor100Lbs   1     9066.1 405920  15860

Step:  AIC=15786.44
WeightFinish ~ VigorInitial + Vigor100Lbs

               Df  Sum of Sq    RSS    AIC
<none>                       396992  15786
- VigorInitial  1     5065   402058  15826
- Vigor100Lbs   1    43088   440080  16124

Call:
lm(formula=WeightFinish ~ VigorInitial+Vigor100Lbs, data=LStockVg.df)

Coefficients:
 (Intercept)   VigorInitial   Vigor100Lbs
     149.228          3.123         8.472
```

From the output of this backward stepwise regression it is now known that the prediction equation that best fits the model is:

```
Y-hat = a + b(x) + b(y)
```

```
WeightFinish = 149.228 + VigorInitial(3.123) + Vigor100Lbs(8.472)
```

Compare the similarity of this prediction equation, generated through the use of a backward stepwise regression with all cases complete (e.g., all cases with missing data were removed from the dataset) to the prior prediction equation that was manually prepared and may have included cases that had missing data.

From a practical viewpoint, the importance of this stepwise regression is that the manager of the livestock finishing operation can influence final weight early on, up to the time weights approach 100 pounds. At 200 pounds and beyond the manager would have less direct influence on attainment of final weight.

As an interesting exercise, use the predict() function to determine the model-based prediction of final weight for an animal with a vigor value of 8.00 for all four vigor measures: VigorInitial, Vigor100Lbs, Vigor200Lbs, and VigorFinish.

```
predict(Fit.Model.WeightFinish.Vigor, list(VigorInitial=8.00,
   Vigor100Lbs=8.00, Vigor200Lbs=8.00, VigorFinish=8.00))
```

```
         1
242.0264
```

Given these parameters, the animal would have a finished weight of 242.0264, which is below the desired finished weight of 250 pounds. Now look at what happens if VigorInitial and Vigor100Lbs are both increased to 8.55 while all other measures remain the same.

```
predict(Fit.Model.WeightFinish.Vigor, list(VigorInitial=8.55,
   Vigor100Lbs=8.55, Vigor200Lbs=8.00, VigorFinish=8.00))
```

```
         1
249.0175
```

Correlation and regression are focused on groups and not specific individuals. Even so, it is interesting to adjust predictors and to use this exercise as a tool in model building for future desired results. This limited exercise may help the livestock finishing operator justify the purchase cost of feeder stock that meet or exceed VigorInitial=8.55 and to also apply management practices to be sure that Vigor100Lbs=8.55. Again, this is merely one example of how biostatistics can contribute to both better knowledge of the biological sciences and also contribute to economics.

7.9 Prepare to Exit, Save, and Later Retrieve This R Session

It is common to prepare R syntax in a separate file, using a simple ASCII text editor. If time permits, experiment with Crimson Editor, Tinn-R, or vim, but there are many other possible selections.

```
getwd()              # Identify the current working directory.
ls()                 # List all objects in the working
                     # directory.
ls.str()             # List all objects, with finite detail.
list.files()         # List files at the PC directory.

save.image("R_Lesson_Correlation-Regression.rdata")

getwd()              # Identify the current working directory.
ls()                 # List all objects in the working
                     # directory.
ls.str()             # List all objects, with finite detail.
list.files()         # List files at the PC directory.

alarm()              # Alarm, notice of upcoming action.
q()                  # Quit this session.
                     # Prepare for Save workspace image? query.
```

Use the R Graphical User Interface (GUI) to load the saved rdata file: **File** and then **Load Workspace**. Otherwise, use the load() function, keying the full pathname, to load the .rdata file and retrieve the session.

Recall, however, that it may be just as useful to simply use a R script file (typically saved as a .txt ASCII-type file) and recreate the analyses and graphics, provided the data files remain available.

Chapter 8
Future Actions and Next Steps

Abstract The purpose of this lesson is to provide a brief summary of this text and to introduce a few topics that should be considered once initial skills in the use of R for biostatistics have been mastered. Along with a few ideas on external resources, contact information for the author is also provided.

Keywords Biostatistics • Code Book • Comma-separated values ASCII file • .csv • Command Line Interface (CLI) • Comprehensive R Archive Network (CRAN) • Data analysis • Descriptive statistics • Graphical User Interface (GUI) • Open source software • R

8.1 Use of This Text

Scientists use empiricism to guide and validate decisions. Precision, orderliness, analysis, and a sound background in statistics are directly associated with informed judgment, decision-making, and the subsequent allocation of human, physical, and fiscal resources – all to improve the human condition. The purpose of this text is to provide an introduction to the use of R software as a platform for analyses related to biostatistics. Data identification, data organization, use of a Code Book, graphical and descriptive portrayal of phenomena, and statistical tests through the use of R are all inherent to this text.

This text serves as an *introduction* to the use of R in biostatistics. This text has specifically been structured to demonstrate the use of R syntax as opposed to the use of a point-and-select Graphical User Interface. This approach, using syntax at the Command Line Interface, gives the most flexibility to the user in the attempt to reach desired outcomes. With even only minimal (if any) experience with programming, the examples in this text empower the user to organize data, prepare graphical images as desired, assess descriptive statistics and measures of central tendency, and complete a wide range of statistical analyses typically found in biostatistics.

Again, this text covered descriptive statistics (e.g., mode, median, mean, SD, range, etc.) and then provided an introduction to some of the leading inferential tests in biostatistics: Student's t-Test, ANOVA, Correlation and Regression. Along with the standard statistical tests, a fair degree of emphasis was placed on the preparation of supporting graphics. Statistical tests are needed and should never be avoided, but when making public presentations graphical images are the media that generate interest and further attention to final outcomes.

8.2 Future Use of R for Biostatistics

The R user community is international and it is growing. R is free and at first the notion of free software may be the compelling consideration for the use of R. Before too long, however, the power and flexibility of R becomes apparent. For those who regularly engage in research, and subsequently associated statistical analyses, the freedom to explore different analyses and to share ideas with like-minded researchers, through the use of user-generated packages, simply cannot be ignored. Again, eventually the power and flexibility of R overrides cost issues for those with advanced needs.

Going beyond the limits of an introductory text, those with further interest in the use of R for biostatistics should consider additional topics such as:

- Data serving as markers for date and time
- Dose-response analysis
- Logistic regression
- Missing data and data imputation
- Nonparametric statistics
- Odds and odds ratios
- Power
- Risk assessment
- Sampling and required sample size
- Survey analysis
- Survival analysis

Small and easy-to-follow confidence-building examples have been used throughout this text. Complexity, often through the use of arguments and functions from external packages, was gradually introduced. The final chapters use a fairly robust approach to the use of R. Even with this gradual approach, the emphasis has been on good programming practices, easy-to-follow actions, and emphasis on graphical presentations, modularity, and syntax reuse.

8.3 External Resources

The Web-page for this text, provided by the publisher, should be a first resource for access to sample sections of selected chapters, access to the .csv datasets, etc. External resources, such as R-specific listserv e-mail discussion groups, should also be considered. Use the many resources provided by the Comprehensive R Archive Network (CRAN, http://cran.us.r-project.org/) also. Of course, what may seem to be countless resources on R are also available by using RSeek (http://rseek.org).

8.4 Contact the Author

Contact the author (Dr. Thomas W. MacFarland, Senior Research Associate and Associate Professor, Nova Southeastern University, Fort Lauderdale, Florida, USA, tommac@nova.edu) if there are any questions about the use of this text, details on the accompanying .csv datasets, or if additional pointers on R for biostatistics are needed. When using e-mail, use a meaningful and descriptive term in the **Subject** header so that the message does not get sent to a SPAM folder or otherwise ignored.

The manufacturer's authorised representative in the EU is Springer Nature Customer Service Centre GmbH, Europaplatz 3, 69115 Heidelberg, Germany. If you have any concerns regarding our products, please contact ProductSafety@springernature.com

Printed and bound by CPI Group (UK) Ltd, Croydon, CR0 4YY

23/03/2026

02076369-0007